HEALTH EMERGENCY PREPAREDNESS AND RESPONSE

Andy dedicates this book to all those who have inspired him and supported him throughout his career. Derek, Tony and Mick have been inspirational in their leadership of emergency planning and response. Their experiences and progressive thinking have constantly been a source of motivation.

Andy's parents Pat and Barry, and partner Stuart have provided unquestionable support in his passion for evidence-based practice and education in emergency preparedness, resilience and response. Their encouragement and endless supply of tea have helped to make this book possible.

Chloe would like to thank family, friends and colleagues for their support and input, and for making the field of health emergency preparedness, resilience and response continuously challenging, interesting and enjoyable.

HEALTH EMERGENCY PREPAREDNESS AND RESPONSE

Edited by

Chloe Sellwood

BSc (Hons), PhD, FRSPH, DipHEP
National Lead Pandemic Influenza, NHS England
Honorary Associate Professor, Health Emergency Preparedness, Resilience and Response, University of Nottingham

and

Andy Wapling

OStJ, MSc, FRSPH, FICPEM
Regional Head of Emergency Preparedness, Resilience and Response, NHS England (South)

CABI

CABI is a trading name of CAB International

CABI
Nosworthy Way
Wallingford
Oxfordshire OX10 8DE
UK

CABI
745 Atlantic Avenue
8th Floor
Boston, MA 02111
USA

Tel: +44 (0)1491 832111
Fax: +44 (0)1491 833508
E-mail: info@cabi.org
Website: www.cabi.org

T: +1 (617)682-9015
E-mail: cabi-nao@cabi.org

© CAB International 2016. All rights reserved. No part of this publication may be reproduced in any form or by any means, electronically, mechanically, by photocopying, recording or otherwise, without the prior permission of the copyright owners.

A catalogue record for this book is available from the British Library, London, UK.

Library of Congress Cataloging-in-Publication Data

Names: Sellwood, Chloe, editor. | Wapling, Andy, editor.
Title: Health emergency preparedness and response / edited by Chloe Sellwood and Andy Wapling.
Description: Wallingford, Oxfordshire, UK ; Boston, MA : CABI, [2016] | Includes bibliographical references and index.
Identifiers: LCCN 2016013450| ISBN 9781780644554 (alk. paper) | ISBN 9781786390431 (epub) | ISBN 9781786390431 (PDF)
Subjects: | MESH: Disaster Planning--organization & administration | Emergencies | Risk Management | Emergency Medical Services | Public Health | Health Policy
Classification: LCC RA645.5 | NLM WA 295 | DDC 363.34/8--dc23 LC record available at
http://lccn.loc.gov/2016013450

ISBN-13: 978 1 78064 455 4

Commissioning editors: Rachel Cutts and Caroline Makepeace
Associate editor: Alexandra Lainsbury
Production editor: Tracy Head

Typeset by SPi, Pondicherry, India.
Printed and bound in the UK by Severn, Gloucester.

Contents

Contributors		vii
About the Editors		ix
Preface		xi
1	Introduction: Why Do We Need to Prepare? *C. Sellwood and A. Wapling*	1
2	The Planning Process *C. Sellwood and A. Wapling*	10
3	Risk Assessment *J. Bush*	18
4	Writing an Emergency Plan *A. Wapling*	30
5	Emergency Planning and Response: Working in Partnership *S. Lewis*	37
6	Interprofessional Working: Understanding Some Emotional Barriers and Unconscious Processes That Might Influence Practice in Group and Team Work *P. Sully*	50
7	Command, Control and Communication *A. Rowe and P. Thorpe*	60
8	Communications During a Health Emergency *J. Cole*	72

9	Psychosocial and Mental Health Care Before, During and After Emergencies, Disasters and Major Incidents *R. Williams and V. Kemp*	82
10	Business Continuity *J. Hebdon*	99
11	Training and Exercising for Emergency Preparedness, Resilience and Response *R. Ellett and A. Wapling*	109
12	Post-incident Follow-up *K. Reddin and G. Macdonald*	126
13	Mass Casualty Incidents *M. Shanahan*	138
14	Preparedness and Response to Pandemics and Other Infectious Disease Emergencies *J.S. Nguyen-Van-Tam and P.M.P. Penttinen*	152
15	CBRN Incidents *R.P. Chilcott and S.M. Wyke*	166
16	A Military Case Study *D. Ross and A. Charnick*	181
17	From Pandemics to Earthquakes: Health and Emergencies in Canterbury, New Zealand *A. Humphrey*	194

Index   211

Contributors

Jacky Bush, MSc DipFM DipHEP, *Corporate Risk Manager, Waitemata District Health Board, New Zealand*

Andrew Charnick, CMCIEH Grad IOSH AMEPS, *Former Defence Specialist Advisor in Environmental Health, Army Health Unit, Former Army Staff College, Surrey, UK*

Professor Robert P. Chilcott, *Head of Toxicology, Department of Pharmacy, University of Hertfordshire, Hatfield, UK*

Jennifer Cole, *Senior Research Fellow, Resilience and Emergency Management, Royal United Services Institute, London, UK*

Rob Ellett, *Former Fire Commander, Humberside Fire and Rescue Service; and Course Director at the Emergency Planning College, Easingwold, Yorkshire, UK*

James Hebdon, *Emergency Preparedness, Resilience and Response Senior Officer, NHS England Central Team, UK*

Dr Alistair Humphrey, *Canterbury Medical Officer of Health, Christchurch, New Zealand*

Verity Kemp, *Director, Healthplanning Ltd; and Associate, Welsh Institute for Health and Social Care, University of South Wales, Pontypridd, UK*

Simon Lewis, *Head of Crisis Response, British Red Cross, London, UK*

Gordon Macdonald, *Director of Studies, Organisational Resilience Programmes, Loughborough University, Loughborough, UK*

Professor Jonathan S. Nguyen-Van-Tam, MBE BMedSci BM BS DM CSci CBiol FFPH FRCPath Hon FFPM FRSPH FSB, *Professor of Health Protection, University of Nottingham School of Medicine, Nottingham, UK*

Dr Pasi M.P. Penttinen, MD MPH PhD, *Head of Disease Programme: Influenza and other Respiratory Viruses, European Centre for Disease Prevention and Control (ECDC), Solna, Sweden*

Dr Karen Reddin, *Strategic Emergency Planning Manager, Public Health England, London, UK*

David Ross, QHP MSc MBBS MRCGP FRCPCH FFPHM FFTM RCPS(Glasg) L/RAMC, *Parkes Professor of Preventive Medicine, Army Health Unit, Former Army Staff College, Surrey, UK*

Anthony Rowe, QPM, *Retired Metropolitan Police Operational Commander, London, UK; and Service Senior Emergency Planning Manager for the London Ambulance Service NHS Trust, London, UK*

Philippa Sully, MSc RN RM RHV CertEd RNT CC Relate, *Honorary Visiting Researcher, Department of Psychology, University of Westminster, London, UK*

Dr Chloe Sellwood, BSc (Hons) PhD FRSPH DipHEP, *National Lead Pandemic Influenza, NHS England, London, UK; and Honorary Associate Professor, Health Emergency Preparedness Resilience and Response, University of Nottingham, Nottingham, UK*

Mike Shanahan, *Head of Special Operations, Yorkshire Ambulance Service NHS Trust, UK*

Peter Thorpe, *Executive Director of the British Columbia Ambulance Service, Canada; previously Head of Olympic Planning for the London Ambulance Service NHS Trust, London, UK*

Andy Wapling, OStJ MSc FRSPH FICPEM, *Regional Head of Emergency Preparedness, Resilience and Response, NHS England (South), UK*

Professor Richard Williams, OBE TD MB ChB FRCPsych FRCPCH DPM DMCC MHSM MIoD, *Emeritus Professor of Mental Health Strategy, Welsh Institute for Health and Social Care, University of South Wales, Pontypridd, UK*

Dr Stacey M. Wyke, *Principal Public Health Scientist, Centre for Radiation, Chemical and Environmental Hazards, Public Health England, Chilton, UK*

About the Editors

Chloe Sellwood, BSc (Hons) PhD FRSPH DipHEP, is the National Lead Pandemic Influenza for National Health Service (NHS) England, within the Emergency Preparedness, Resilience and Response (EPRR) Team. She leads NHS England internal pandemic influenza preparedness as a subject matter expert and is coordinating national pandemic preparedness across the NHS, with a specific focus on London. Her experience in pandemic influenza ranges from local to international levels and encompasses scientific, strategic and operational aspects, in both preparedness and response. She spent over 7 years at the Health Protection Agency, including 3 years as the Senior Scientist and Coordinator of the Pandemic Influenza Office. In 2008 she joined NHS London (the then strategic health authority for London) as the Pandemic Influenza Resilience Manager and was heavily involved in the response to the swine flu pandemic. In 2010 she assumed the additional role of 2012 Health Resilience for the NHS across London for the Olympic and Paralympic Games. Since autumn 2014 she assumed the strategic leadership for NHS Ebola preparedness in London. She is the co-editor of, and a contributing author to, two textbooks on pandemic influenza, as well as many other articles and papers on influenza resilience. She has worked with the World Health Organization and the European Centre for Disease Prevention and Control on international consultations, as well as on secondment to the Department of Health (England) Pandemic Influenza Preparedness Programme.

Andy Wapling, OStJ MSc FRSPH FICPEM, is the Head of Emergency Preparedness, Resilience and Response for NHS England South Region. Andy provides strategic leadership and assurance across providers and commissioners of NHS-funded care for a quarter of the UK population. Andy has been involved in emergency preparedness for over 20 years where he has worked in private, voluntary and public sector organizations. He has been directly involved in the response to many emergencies including the

Paddington rail disaster 1999, Soho bombing 1999, 7th July London bombings 2005, Yorkshire flooding 2007, pandemic influenza 2009 and many more. Andy obtained a Master's degree in Civil Emergency Management in 2005 at City University London and continues to support others in their education as a visiting lecturer at Loughborough University on the Diploma of Health Emergency Preparedness Resilience and Response. Andy maintains his connection with the voluntary sector and is the National Service Delivery Adviser for St John Ambulance. As a part of this role Andy is responsible in providing the charity with advice on national emergency preparedness policy. Andy was admitted into the Most Venerable Order of St John of Jerusalem in 2001 and further promoted to Officer of the Order in 2012.

Preface

Health emergency preparedness, resilience and response (EPRR) is a rich and varied field. Any major incident or emergency is likely to have some impact on the health of those involved in the response or affected by the incident itself. Individuals from all manner of organizations and backgrounds across the public, private and voluntary sector are increasingly becoming involved in this work.

We originally conceived the idea for this book following the 2009/10 influenza pandemic. However there have been many delays and restructures to its format because many of the contributing authors have been involved, in one way or another, in responding to a number of subsequent major incidents in the UK and overseas. These have included events like terrorist attacks, transport incidents, wide-scale flooding and the Ebola outbreak in West Africa that started in 2014 (the biggest infectious disease outbreak since the influenza pandemic).

This book covers a range of aspects when planning for health emergencies and is illustrated throughout with a number of real-life case studies. We have drawn on our rich network of colleagues and friends from across the world to contribute to this and are very grateful to them for their support and patience through this project. The style of the chapters ranges from academic to conversational, as well as reflecting local language, terms and descriptors. This represents the background and experiences of the authors, and we feel this adds to the colour of the book. The book is prepared so that chapters can be read in isolation and without reading in any particular order.

We hope you find this a helpful addition to the health EPRR literature.

Chloe Sellwood and **Andy Wapling**
June 2016

1 Introduction: Why Do We Need to Prepare?

C. Sellwood[1] and A. Wapling[2]

[1]National Lead Pandemic Influenza, NHS England, London, UK; and Honorary Associate Professor, Health Emergency Preparedness, Resilience and Response, University of Nottingham, Nottingham, UK
[2]Regional Head of Emergency Preparedness, Resilience and Response, NHS England (South), UK

Key Questions

- What is the key underpinning legislation for emergency preparedness, resilience and response (EPRR) in the UK?
- What are the main categories of major incidents and emergencies?
- Who benefits from EPRR processes being embedded in health organizations?
- What could be the repercussions of health organizations not undertaking or engaging in emergency preparedness activity?
- What is the Sendai Framework and why is it relevant to health emergency preparedness?

1.1 Introduction

Incidents and emergencies, by their nature, can occur at any time and in any place. Man-made, accidental or naturally occurring, these can pose significant threats to the health of the population. From earthquakes to terrorism there is a responsibility for communities to have arrangements in place to preserve life, prevent deterioration and promote recovery.

Some of the first questions to consider with regard to emergency preparedness, resilience and response (EPRR) in the health sector is 'why do we need to plan?' and 'why can't we just use existing systems and processes?' This book attempts to answer these questions through the subsequent chapters.

Preparing for unique, rare or extreme events results in many benefits to those affected by the emergency, to the responders and to the effective running of organizations. It is important that the response to any major incident is through a structured and coordinated framework within which responders can operate safely and effectively. This is most effective when it reflects existing systems as new processes at the time of an incident response could result in unnecessary suffering and potentially lives being lost. Staff would also face unnecessary stress and resources could be wasted as people and organizations try to respond in an uncoordinated or haphazard manner. This chapter discusses some of the benefits of planning, as well as why we prepare for emergencies.

1.2 Legislative Setting

The UK Civil Contingencies Act (2004) defines an emergency as:

> an event or situation which threatens serious damage to human welfare in a place in the UK, the environment of a place in the UK, or war or terrorism which threatens serious damage to the security of the UK.

The Cabinet Office National Risk Register currently identifies a number of threats and hazards to the UK, as illustrated in Fig. 1.1.

The terminology used to describe such events is varied and includes 'emergencies', 'major incidents' and 'civil emergency', among others. Equally, they can range in size and impact from something affecting a village or town (such as localized flooding), to something affecting a discrete population (such as a major transport incident or release of a chemical), to something affecting whole countries or even the world (such as an outbreak of an infectious disease like Ebola or pandemic influenza).

Rightfully health organizations are involved in planning for and responding to more and more scenarios – both health-specific events as well as the health impacts of other emergencies. These include big-bang events such as explosions, cloud-on-the-horizon events such as the plume from a volcanic eruption, and rising-tide events such as pandemics (Table 1.1).

The Civil Contingencies Act (2004) places statutory duties on many organizations in the UK to prepare for and respond to major incidents and emergencies. This was passed into law following a number of major incidents in the UK and overseas. The incidents ranged in size, location and cause, but all affected people and communities. Many of the reviews of these incidents identified common areas for improvement such as better joint working between responding organizations, better capabilities and equipment, and better communication processes.

1.3 Health Service and Systems Preparedness

To paraphrase one of the authors of a later chapter in this book: there are no health emergencies; all emergencies have health aspects. It is increasingly important that health organizations across the breadth of providers and commissioners, in public, private and voluntary sectors, undertake and engage in EPRR activities.

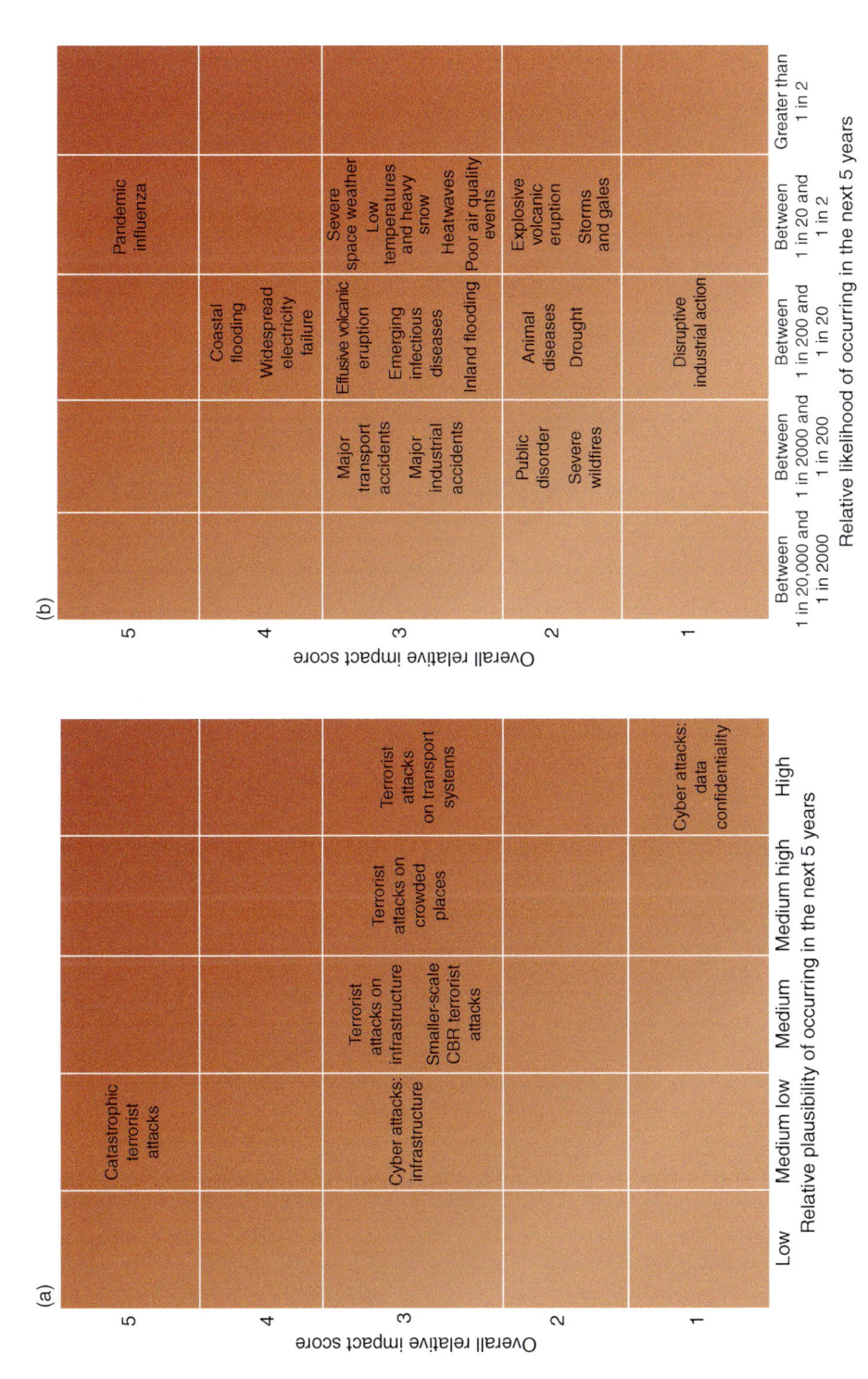

Fig. 1.1. Hazards and threats to the UK as identified in the Cabinet Office National Risk Register 2015: (a) risks of terrorist and other malicious attacks (CBR, chemical/biological/radiation); (b) other risks. (Reproduced with permission from www.gov.uk/government/uploads/system/uploads/attachment_data/file/419549/20150331_2015-NRR-WA_Final.pdf under the Open Government Licence (www.nationalarchives.gov.uk/doc/open-government-licence/version/3/).)

Table 1.1. Types of major incidents and emergencies.

Type	Example
Big bang	An explosion or major transport incident
Cloud on the horizon	A significant chemical or nuclear release developing elsewhere and needing preparatory action
Rising tide	Epidemic or pandemic of infectious disease, or a capacity/staffing crisis (e.g. industrial action)
Headline news	Public or media alarm about an actual or impending situation (e.g. the MMR (measles, mumps, rubella) vaccine issues)
Internal incidents	Utility or equipment failure, fire, hospital-acquired infections, violent crime
CBRN(e)	Deliberate (criminal intent) release of chemical, biological, radioactive or nuclear materials or explosive device
HAZMAT	Incident involving hazardous materials (typically non-malicious)
Mass casualties/fatalities	Incident resulting in significant numbers of casualties or fatalities that would potentially overwhelm the capacity of a single organization to cope

Many organizations in the UK have a legal or statutory obligation to prepare for and respond to major incidents. In addition, it is good practice for all primary, secondary and tertiary health care providers to undertake business continuity management (BCM) processes and to engage with their local communities and partner organizations to ensure they can continue to deliver services during a disruption, or respond to the external challenges of a major incident.

Guidance in the UK for the NHS on EPRR has been led by NHS England since 2013. An overarching framework and annual assurance process, with periodic specialist subject deep-dive assessments, is helping to ensure that EPRR activity is embedded within organizations and accorded due attention and status (https://www.england.nhs.uk/ourwork/eprr/gf/).

In many incident scenarios, health organizations can face a double challenge of both responding to the incident (e.g. treating increased numbers of patients with broken hips or hypothermia during extended periods of severe cold weather) as well as facing the complication of reduced staffing (e.g. due to transport disruption caused by heavy snowfall).

Additionally, health care settings themselves can become the scene of a major incident – such as a fire or flood – which means that the responders themselves equally become entangled in the incident as 'victims'. In 2008/09 London experienced five hospital fires across the capital that required the evacuation of part or all of the building.

These events proved that with good teamwork, leadership and planning, a safe and successful evacuation of a health care facility is achievable. London's experiences during 2008/09 demonstrate the critical importance of being prepared for all emergencies.

1.4 Planning in Partnership

We live in an increasingly complex and intertwined society. It is rare that a single, individual organization will be able to respond effectively in isolation to a major incident. There is increasing scrutiny by the public and politicians through instant 24/7 media access. The benefits of getting preparedness and response right are clearly that lives are saved and normality is restored promptly. However, if we get it wrong, reputations can be ruined, trust lost and the financial consequences can be severe.

For some organizations the greatest risk could be the loss of reputation or confidence, which is just as important for health care organizations as it is for finance and retail businesses. For the health service, this could be the risk of failure to provide emergency and life-saving services. EPRR and business continuity processes (Chapter 10, this volume) will help to identify reputational risks if an organization fails to respond to a major incident.

1.5 The Benefits of Planning

There are clear benefits to responders and the public in organizations and individuals having prepared for a range of possible scenarios. In all cases the patient must be at the centre of planning and response arrangements and due consideration must be given to the health and safety of responders.

Failure to plan in advance could mean that lives are unnecessarily lost or negatively impacted. This could be people immediately injured in an incident such as a major transport collision, those involved in the response who could be exposed to a dangerous substance (e.g. when responding to a chemical, biological, radiation or nuclear (CBRN) incident) or through psychosocial trauma some weeks, months or years after an incident.

From an organizational perspective, businesses could be damaged through loss of reputational status, loss of business or legal action. These are all increasingly real concerns and have occurred to a number of health care organizations in a range of incidents both within and outside the field of EPRR.

It is essential that the process of preparing to respond to major incidents is embedded in organizational structures, is regularly reviewed and considers all possibilities. One criticism that has been laid at the field of emergency preparedness is that of 'preparing to respond to the previous disaster'. Thus horizon scanning is an essential component of any robust EPRR strategy.

Many lessons have been identified from the response to the outbreak of Ebola virus that started in West Africa in 2014. It is important that these lessons are learned and applied in response to future outbreaks of Ebola or Ebola-like pathogens; however, it is equally important that planning for emerging infectious diseases continues to consider a range of pathogens, vectors and clinical presentations. While it is certain that another pathogen (be it a virus, bacterium or other agent) will emerge from an unknown reservoir at an unknown time in an unknown location into human populations, there is no certainty around how it will spread, who might be susceptible, how the disease will manifest and what treatment will be required. Plans must therefore remain flexible to respond to a range of circumstances.

1.6 The Global Setting

The Sendai Framework for Disaster Risk Reduction 2015–30 (Sendai Framework) is being delivered by the United Nations Office for Disaster Risk Reduction (UNISDR) following endorsement by the UN General Assembly and adoption by UN Member States in March 2015. It is a 15-year, voluntary and non-binding agreement which, while recognizing that the Member State has the primary role to reduce disaster risk, identifies that the responsibility for disaster risk reduction in preparedness and response should be shared with other stakeholders. This is a key principle which is reflected throughout the chapters in this book.

The Sendai Framework has seven targets and four priorities (Table 1.2) towards preventing new risks and reducing existing risks. This overall aim has been summarized as:

> the substantial reduction of disaster risk and losses in lives, livelihoods and health and in the economic, physical, social, cultural and environmental assets of persons, businesses, communities and countries.

The Framework includes specific references to health impacts, such as mortality, morbidity, population displacement and economic repercussions. There are focused sections considering health infrastructure, health innovation and technology, health system resilience, disaster risk management for health, access to health care services, life-threatening and chronic diseases, metal health and stockpiling. Many of these elements are discussed in more detail throughout this book, particularly through the scenarios and case studies.

1.7 The Rest of this Book

This book includes contributions from many different authors with different backgrounds from across the world, across sectors and across experiences. Included are experienced practitioners in health EPRR, BCM and communications. They come from the public, private and voluntary sectors, academia and the military. There are a number of leading global experts in subjects such as infectious diseases and CBRN threats, and experienced academics in the fields of interagency interoperability and psychosocial support. This is reflected in the varied style of the chapters, and a conscious decision not to consistently use formal referencing has resulted in an accessible narrative for readers of all levels of experience, which is supported with suggested further reading that the authors have identified to add further context and detail to their chapters.

Table 1.2. The Sendai Framework targets and priorities.

Global targets	Priorities
1. Substantially *reduce global disaster mortality* by 2030, aiming to lower the average per 100,000 global mortality rate in the decade 2020–2030 compared with the period 2005–2015 **2.** Substantially *reduce the number of affected people* globally by 2030, aiming to lower the average global figure per 100,000 in the decade 2020–2030 compared with the period 2005–2015 **3.** *Reduce direct disaster economic* loss in relation to global gross domestic product (GDP) by 2030 **4.** Substantially *reduce disaster damage to critical infrastructure and disruption of basic services*, among them *health* and educational facilities, including through developing their resilience by 2030 **5.** Substantially *increase the number of countries* with national and local *disaster risk reduction strategies* by 2020 **6.** Substantially *enhance international cooperation to developing countries* through adequate and sustainable support to complement their national actions for implementation of this Framework by 2030 **7.** Substantially *increase the availability of and access to* multi-hazard early warning systems and disaster risk *information and assessments* to the people by 2030	*Priority 1. Understanding disaster risk.* Disaster risk management should be based on an understanding of disaster risk in all its dimensions of vulnerability, capacity, exposure of persons and assets, hazard characteristics and the environment. Such knowledge can be used for risk assessment, prevention, mitigation, preparedness and response *Priority 2. Strengthening disaster risk governance to manage disaster risk.* Disaster risk governance at the national, regional and global levels is very important for prevention, mitigation, preparedness, response, recovery and rehabilitation. It fosters collaboration and partnership *Priority 3. Investing in disaster risk reduction for resilience.* Public and private investment in disaster risk prevention and reduction through structural and non-structural measures is essential to enhance the economic, social, health and cultural resilience of persons, communities, countries and their assets, as well as the environment *Priority 4. Enhancing disaster preparedness for effective response and to 'Build Back Better' in recovery, rehabilitation and reconstruction.* The growth of disaster risk means there is a need to strengthen disaster preparedness for response, take action in anticipation of events, and ensure capacities are in place for effective response and recovery at all levels. The recovery, rehabilitation and reconstruction phase is a critical opportunity to build back better, including through integrating disaster risk reduction into development measures

The following chapters include:

- a summary of the planning process;
- a discussion on the process of risk assessment;
- how to write a plan;
- the benefits of planning and responding in partnership with other organizations;

- aspects of command and control;
- the key communications elements of planning for and responding to emergencies;
- dealing with the personal impact of emergencies on patients and staff through psychosocial support;
- the relevance of BCM;
- the importance of training, testing and exercising response arrangements; and
- post-incident follow-up.

These technical aspects are then further elaborated on and illustrated through a series of case studies describing the preparedness for and response to:

- mass casualty events;
- infectious disease outbreaks, epidemics and pandemics;
- CBRN events;
- the role of the military in response; and
- particular challenges relevant to earthquakes.

This book does not attempt to provide a detailed 'how to' guide to health emergency planning and response; instead it aims to provide a series of informative descriptions of key elements that are underpinned by real-life examples. The wealth of experience from the authors is easy to see when reading the chapters and while some terms may not all be instantly familiar to all readers, the principles can easily be adopted, adapted and applied.

> **Key Answers**
>
> - The Civil Contingencies Act 2004 is the key piece of legislation underpinning EPRR guidance in the UK.
> - The main types of major incident are labelled as: big bang, cloud on the horizon, rising tide, headline news, internal incidents, CBRN(e), HAZMAT and mass casualty/fatality.
> - Everyone benefits from health organizations having embedded EPRR processes; this includes staff, partners, patients, members of the wider public and the organization itself.
> - If health organizations do not engage in EPRR activities, lives could be unnecessarily lost or damaged, the reputation and trust in organizations could be lost, or individuals could be found criminally liable for not meeting statutory obligations.
> - The Sendai Framework is a 15-year agreement which identifies that the responsibility for disaster risk reduction in preparedness and response is a partnership responsibility; it specifically describes a number of issues relevant to health care settings.

Further Reading

Cabinet Office (2006) Emergency preparedness: Guidance on part 1 of the Civil Contingencies Act 2004, its associated regulations and non-statutory arrangements. Available at: www.gov.uk/government/publications/emergency-preparedness (accessed 10 October 2015).

Cabinet Office (2011) Chapter 5 (Emergency Planning) of Emergency Preparedness, Revised Edition. Available at: https://www.gov.uk/government/uploads/system/uploads/attachment_data/file/61028/Emergency_Preparedness_chapter5_amends_21112011.pdf (accessed 13 July 2016).

Dzau, V.J. and Rodin, J. (2015) Creating a global health risk framework. *New England Journal of Medicine* 373, 991–993.

Gates, B. (2015) The next epidemic – lessons from Ebola. *New England Journal of Medicine* 372, 1381–1384.

London Emergency Services Liaison Panel (2015) LESLP Manual – Ninth Edition. Available at: www.leslp.gov.uk (accessed 9 February 2016).

NHS England (2015) NHS England Core Standards for Emergency Preparedness, Resilience and Response (EPRR). Available at: www.england.nhs.uk/ourwork/eprr/gf/#core (accessed 14 December 2015).

NHS England (2015) NHS England Emergency Preparedness, Resilience and Response Framework. Available at: www.england.nhs.uk/ourwork/eprr/gf/#preparedness (accessed 14 December 2015).

NHS London (2009) Review of five London hospital fires and their management, January 2008–February 2009. Available at: www.preventionweb.net/files/13954_reviewoflondonhospitalfires1.pdf (accessed 30 December 2015).

NHS London (2010) Review of the London health system response to the 2009/10 influenza A/H1N1 pandemic. Available at: www.webarchive.org.uk/wayback/archive/20130304124415/http://www.london.nhs.uk/webfiles/Emergency%20planning%20docs/NHSL_FLU_REPORT_WEB.pdf (accessed 30 December 2015).

United Nations Office for Disaster and Risk Reduction (2016) Fact sheet: health in the context of the Sendai framework for disaster risk reduction. Available at: www.unisdr.org/we/inform/publications/46621 (accessed 10 February 2016).

United Nations Office for Disaster and Risk Reduction (2016) Sendai Framework for Disaster Risk Reduction. Available at: www.unisdr.org/we/coordinate/sendai-framework (accessed 10 February 2016).

2 The Planning Process

C. Sellwood[1] and A. Wapling[2]

[1]*National Lead Pandemic Influenza, NHS England, London, UK; and Honorary Associate Professor, Health Emergency Preparedness, Resilience and Response, University of Nottingham, Nottingham, UK*
[2]*Regional Head of Emergency Preparedness, Resilience and Response, NHS England (South), UK*

> **Key Questions**
> - What is the process that should be adopted to undertake effective emergency preparedness?
> - What are the stages that can be followed within this process?
> - How often should the process be revisited?

2.1 Introduction

One of the greatest benefits of the emergency planning process is not necessarily the plan that is delivered at the end, but the actual process of developing the plan. Additionally, this often helpfully identifies who to speak to, how to communicate with them and, perversely, what NOT to do, all of which are essential aspects of responding promptly, appropriately and safely to a major incident. This chapter, which is based on a wider discussion on emergency preparedness and business continuity in *Pandemic Influenza*, 2nd edn (J. Van-Tam and C. Sellwood (eds) 2013), discusses the planning process, using the integrated emergency management model as a basis.

2.2 Response Planning

While there are a few different planning models available, all incorporate some common key principles. These include the need for risk-based planning and proportionality; integration with existing systems, processes and partner organizations; and a means to verify the plan.

One of the most effective planning models is that advocated by the UK Cabinet Office, which, in line with the concept of integrated emergency management, provides a continual cycle of planning, review and revision. Good practice emergency planning follows a continual eight-point emergency preparedness cycle, which is repeated at intervals. There is one addition that is often made to this model and that is an initial action to 'assess'. It is important to fully understand the scientific and technical information about the scenario before a specific risk assessment can be undertaken (Fig. 2.1). The following sections discuss each of these elements in more detail.

2.2.1 Assessing the scenario

In order to accurately assess the risk and impact of any scenario, good baseline knowledge of the situation is required. Without this, the impact could be under- or over-assessed and consequently the response would be insufficient or an overreaction.

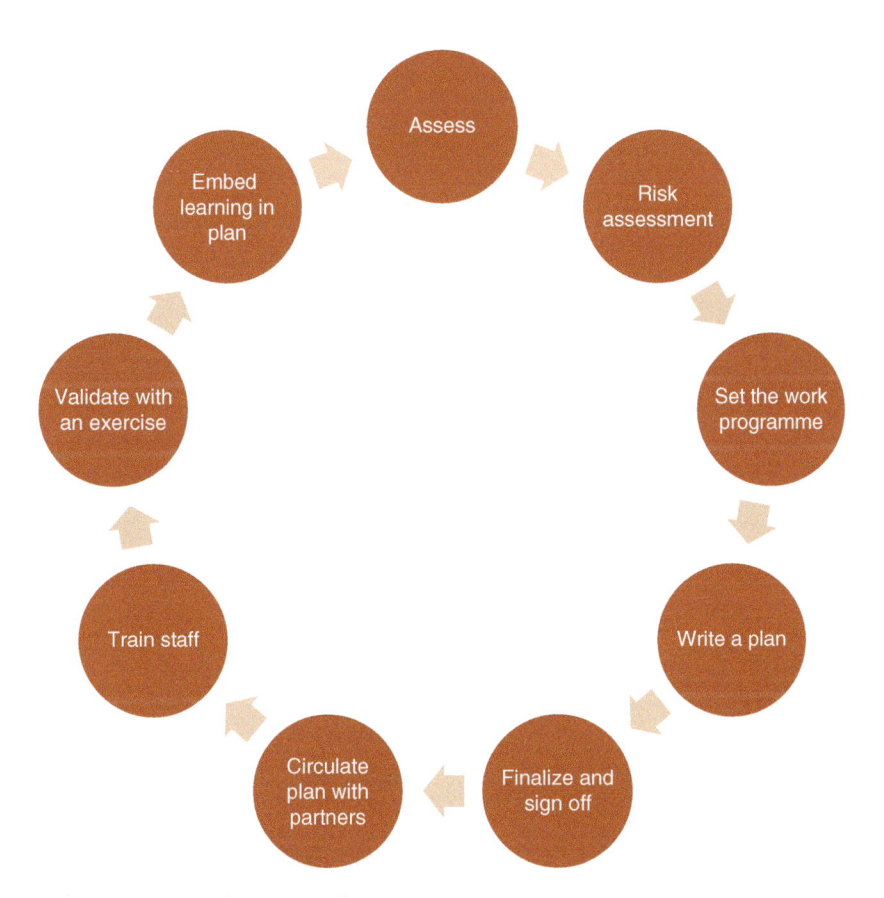

Fig. 2.1. The emergency planning cycle.

Planners don't need to be an expert in every scenario but a good basic understanding of the impact that each scenario could have is important. This is especially true with some of the newly recognized threats and hazards (Box 2.1), such as marauding terrorist attacks, space weather and emerging infections such as Ebola (Chapter 14, this volume).

Early risk assessments of Ebola during late 2014 and early 2015 were not always based on scientific fact. There was misunderstanding by people at all levels (from patient-facing staff right through to senior governments) about basic aspects of the virus and illness, which consequently resulted in inappropriate risk assessments. For example, there was a prevalent misunderstanding by many that the virus was airborne and as such as easy to catch as influenza; whereas in actuality it is spread through close contact with bodily fluids. Until this misconception was recognized and addressed, many of the developing plans were inappropriate and did not accurately reflect the true scenario. Understanding how a virus is spread will affect delivery of interventions such as treatment and vaccines, while an understanding of the likely clinical presentations will affect which part of the health service might experience the biggest impact and so can facilitate directed planning.

The scientific evidence base around hazards and threats is always evolving and developing, and as such the aspect of 'assessing' the risk should be revisited on a regular basis.

2.2.2 Risk assessment

Risk assessment is covered in much greater detail in Chapter 3; however, in summary, assessing the risk of a potential hazard or threat is essential to robust emergency preparedness and response. Risk assessment needs to be based on an understanding of the scenario and should be both proactive and reactive, systematic and dynamic.

At its most basic, risk assessment is the process of assessing the impact of a specific scenario on the health of a population against the likelihood of it occurring. For example, a meteor hitting the earth will have a huge impact on health, but the likelihood is very slim. Conversely, the impact of the common cold on population heath is minimal but the likelihood is substantial.

Box 2.1. Hazard versus threat

- *Hazard*: Accidental or naturally occurring (i.e. non-malicious) event or situation with the potential to cause death or physical or psychological harm, damage or losses to property, and/or disruption to the environment and/or to economic, social and political structures.
- *Threat*: Intent and capacity to cause loss of life or create adverse consequences to human welfare (including property and the supply of *essential services* and commodities), the environment or security.

2.2.3 Establishing and delivering the plan through a work programme

A good understanding of the scenarios and a comprehensive risk assessment will give rise to the priority order for addressing the challenges. Thus the greatest risks should be given the most urgent attention and lower risks addressed in slower time. For example, the risk of staff unavailability during a pandemic might be assessed as higher than the risk of running out of non-pharmaceutical countermeasures and so should have more attention directed to it.

However, planning should not be a linear process and it may be necessary to have a number of projects underway in parallel; indeed, this will be the best way to approach interrelated risks. Additionally, it is important that work plans are flexible and enable reassessment and reprioritization of different work streams. Health emergency, preparedness and response is rarely a static field.

New hazards or threats may be identified (e.g. space weather is something that is becoming increasingly recognized) which redirect attention from established programmes of work that perhaps are perceived as less attractive, interesting or exciting. However, it is essential that something which is topical but perhaps unlikely to have a major impact (or of a low likelihood of occurrence) does not adversely distract resources from established programmes that are planning for scenarios of greater likelihood or greater impact.

Pandemic preparedness is perhaps one area that suffers from this. In the UK it is top of the national risk register for both impact and likelihood; however, planning and preparedness for a future pandemic is often derailed by attention being redirected to 'hot topics' such as the Ebola outbreak in West Africa or the continuing situation with Middle East respiratory syndrome corona virus (MERS-CoV) in Saudi Arabia.

It is therefore important that a comprehensive and realistic programme of planning activity is developed using the information gained from the *assess* and *risk assessment* processes. This programme can helpfully be shared with and agreed in association with partner organizations; planning together will make responding easier and more productive (Chapters 3 and 5, this volume).

2.2.4 Writing a plan

Once a full and comprehensive work programme is in place, it is possible to start developing the plan itself. The end product must be accessible to any reader and easy to use in a real response, without the need for the author to stand alongside and explain elements. It is important to recognize from the outset that very few responders will be interested in the depth of scientific and background knowledge that the author has gleaned through the assessment and risk assessment processes. Even when staff outside the field of health emergency preparedness, resilience and response (EPRR) are actively engaged in the process, few are likely to read the end product from cover to cover in advance of needing it in a response.

The strength of a good plan is that it is structured, succinct, brief, easy to navigate, and provides sufficient operational information to be workable in the event that it is deployed. Additionally, it is important not to 'over plan' or structure the response too tightly. It is impossible to accurately predict the exact course of events for each emergency, therefore plans must be flexible and adaptable. They need to provide sufficient detail to inform the response, but not prescript a response such that it then doesn't fit the emergency. A lack of flexibility within a plan could mean that responders feel constrained or that they start to improvise a response, abandoning aspects that are relevant and appropriate. Practically, one useful solution to address this is through the development of 'frameworks'.

There should be enough background so that the responders have sufficient information to make informed decisions. It should include a command and control framework to manage the response, and a sufficient amount of operational options that the responders can choose which to use depending on the incident and the issues it presents. This is discussed more in Chapter 4.

2.2.5 Senior engagement and sign-off

Emergency preparedness and response is a statutory obligation for the majority of large health care providers in the UK, but regardless of this it is clearly sensible for organizations to be prepared to respond to a major incident, surge in demand or increase in patients, or disruption to service delivery. In support of this, NHS organizations all have an Accountable Emergency Officer (AEO) who is the board-level champion for EPRR. Senior sign-off at board level or equivalent is an important way of demonstrating organizational ownership of the plan. Plans must be owned by organizations and their senior officials, not just the emergency planner or plan writer.

2.2.6 Partnership engagement

Many major incidents and emergencies require a multi-agency response. This is obvious for something such as extreme weather where there is disruption to transport services, interruption to power supply or displaced persons, where local authorities, emergency services and the voluntary sector (as well as transport and utilities providers) may all be involved with the same individuals affected by the incident.

Even scenarios that are often thought of as purely health emergencies – such as an outbreak of infectious disease – will often see other agencies being impacted by the event or getting involved in the response. For example, in the event of an influenza pandemic, no one organization would be able to respond to a pandemic in isolation. The impact could be far reaching into all communities and in turn require a response from a number of community-facing organizations.

Plans should be shared so that all partners know what each other is planning to do and when, so as to inform their own response. Gaps and areas of potential conflict can also then be identified quickly and addressed. This could be taken a step further through the development of joint interagency plans (Chapters 5 and 6, this volume).

2.2.7 Staff training

Training people to respond to incidents and emergencies is of fundamental importance. Some organizations (or departments within organizations) are geared to responding to routine, everyday challenges by following usual business practices – such as ambulance services or emergency departments. However, very few people or organizations will regularly respond to major incidents.

In some settings there can be an expectation that individuals and organizations will just be able to respond to a challenging event. However, this demands practice and if staff are to respond effectively to an emergency they require the knowledge, tools and skills to do so.

Once the plan is complete a training needs analysis (TNA) should be undertaken to ensure that all relevant staff have the skills and abilities to undertake their role in an emergency. It would be negligent to place people in a position that they were not trained for, not just to that member of staff but to the communities that they serve. From the health perspective, this isn't just around clinical skills, but also elements of leadership and media training.

Training must be regular, routine and ongoing, to ensure skills are maintained. Not only do staff change jobs and organizations, but if the skills are not used on a regular basis then they are soon forgotten. Therefore, an accurate list of who has had what training and when must be kept and a cycle of regular updates must form part of the training strategy. It is also sensible to maintain the ability to rapidly roll out additional emergency response training if such skills are required. There is more information about training in Chapter 11.

2.2.8 Exercising the plan for validation

Every plan should be validated through a formal test or exercise before it is deemed fit for purpose and signed off. Without this, a plan would not stand up to any formal scrutiny, such as a public enquiry, following activation in an emergency response. One of the most effective ways to validate a plan is through an exercise. This also provides the ability to test concepts, strategies and solutions in a safe environment and to make mistakes, without participants feeling scrutinized or threatened.

Exercises help staff and organizations walk through plans and identify any issues that require solutions. These useful activities help identify all the little things that are not thought about in the planning stage, but that come to light when the system is tested. Exercises need to be held regularly so that plans are up to date and relevant to the organization. Exercises are discussed in more detail in Chapter 11.

2.2.9 Ensuring learning is incorporated into plan revisions

Before restarting the whole cycle of preparedness, it is important to embed learning from issues identified in exercises or real-life activation back into the plan to inform practice. Following significant incidents and any subsequent review or legal processes, organizations are often asked if lessons identified in exercises informed the plan and response. Many organizations have been found wanting when the lessons have not been embedded into their plan and thus their response; and legal processes have in turn ruled these to be forms of corporate negligence.

2.3 Summary

It is important that emergency response plans are based on a sound understanding of the most up-to-date evidence base, accurately reflect a realistic risk assessment, have senior-level support within organizations, and are tested and exercised with staff who are appropriately trained to respond.

Perhaps the most effective plans are those based on business-as-usual processes that are flexible and can be adapted to a rapidly evolving situation or to an unexpected incident.

Emergency preparedness is a continual process, and is much more than just writing a plan and leaving it on a shelf. It is important that plans are regularly revisited and kept current. Knowledge of a hazard or threat will change as we learn more or as new risks are identified. Additionally, organizations change, people change and indeed populations change. An out-of-date plan is perhaps a greater risk than not having a plan.

Robust and embedded emergency preparedness arrangements are essential components of an organization's ability to respond to a range of challenges from flooded premises, industrial action and loss of IT networks, through to global incidents such as a pandemic or terrorist attacks. Through thoroughly understanding the threat or hazard and undertaking the process of reviewing and writing plans, coupled with training, testing and exercising, organizations are able to respond to the best of their ability, with the best possible outcome for their population, partners and other stakeholders.

> **Key Answers**
>
> - Responding to emergencies requires structure and process in order that it is effective to those caught up in the emergency and supportive to those enacting and delivering the response. This also applies to the preparedness process. A good and effective process provides a framework for planners to follow so as to ensure the best possible response when it is required.
> - The emergency planning cycle is a framework that should be applied in order to support the emergency planner in delivering effective arrangements. This cycle includes assessing the science and evidence in relation to a particular scenario that is then assessed in terms of its risk of occurrence and priority with which it should be addressed. Following this a work plan can be set so that high-priority risks receive the most attention. Once the work plan is set arrangements need to be established and a plan written to encapsulate them, so that staff responding to an emergency do so in an informed and structured manner. The plan will require finalizing and signing off so that the organization at its highest level owns the plan. Plans should then be circulated with relevant partners, so that expectations are overt and shared. A training needs analysis then needs to be completed and the relevant training delivered to responders. It should then be tested and validated by an exercise so that gaps can be identified, rectified and embedded, before the plan is used in a real emergency.
> - The emergency planning process is described as a cycle for good reason. Plans that are written and left on a shelf for years are unlikely to be fit for purpose if the cycle is not revisited on a regular basis. Our understanding of science advances, risk profiles change, staff change, organizations change – all of which influences the ability of a plan to be accurate, up to date and therefore safe to be applied. Therefore, the cycle should be reapplied regularly to all emergency arrangements.

Further Reading

Cabinet Office (2006) Emergency preparedness: Guidance on part 1 of the Civil Contingencies Act 2004, its associated regulations and non-statutory arrangement. Available at: www.gov.uk/government/publications/emergency-preparedness (accessed 10 December 2015).

Wapling, A. and Sellwood, C. (2013) Emergency preparedness and business continuity. In: Van-Tam, J. and Sellwood, C. (eds) *Pandemic Influenza*, 2nd edn. CAB International, Wallingford, UK, pp. 88–96.

3   Risk Assessment

J. BUSH

Corporate Risk Manager, Waitemata District Health Board, New Zealand

---

Key Questions

- What is a risk?
- How are risks assessed and quantified in relation to emergency preparedness?
- What decisions are available once the risk has been quantified?
- How does risk information influence the emergency preparedness work plan?

3.1   What is Risk and Risk Management?

There are many variations of the definition of risk; however, they all have a common theme to them, which is that risk is measured in terms of consequence (impact) and likelihood (chance/probability) of an event occurring. Risk can be described as the uncertainty of outcome, be it positive (an opportunity) or negative (a threat), if an event occurs.

Once there is an understanding of what risk is, there is a need to understand what risk management is and why it should be done. Risk management is not a dark art; it is something everyone does in both work and personal lives, often without realizing it (e.g. the continual risk-based decision making that takes place when travelling to work). Individuals make judgements about risks, actions and safety, and decide on the level of risk they (or the organization) is willing to accept for a perceived gain. Risk management is the process that enables an organization to: understand what risks it faces; understand what events or hazards might cause harm to individuals or the organization; assess the risk; and identify existing and/or additional controls that are needed to either prevent the risk event(s) or mitigate the impact should the risk event(s) materialize.

The process for assessing risk is a key element in achieving an understanding of risks faced by an organization and its staff (or, in terms of emergency planning, the community), how they are being managed and prioritized, and what more needs to be done. Understanding and managing the risks an organization or community faces makes for a successful organization and a safe community. Risk management is therefore the process of identifying, assessing, analysing, evaluating and managing all potential risks. Risk management, however, is not about managing issues, which are events or situations that have already happened, have impacted or are currently impacting on the service and so are causing problems in doing business now. Issues, therefore, are topical and do not necessarily correlate to a risk occurring.

In order for risk management to be successful within an organization, it must be integrated into its culture, be underpinned by a shared organizational vision and have senior management commitment, which is supported by an effective policy. There also needs to be a good organization-wide understanding, often achieved through effective training. Successful risk management will support effective risk identification and the development of necessary plans, such as emergency preparedness plans, to mitigate the risk or ensure an effective response should the risk materialize.

The ability to identify and manage risk is an essential leadership skill and a key element in effective governance; it enables organizations to demonstrate that they are doing their reasonable best to control significant risks or prepare for their realization proactively, rather than acting reactively to adverse events. It can, therefore, be seen that good risk management equals good management: it reduces the probability of failure, increases the probability of success and reduces volatility, meaning that there are fewer surprises for the organization. Risk management underpins many business processes, including emergency planning and business continuity planning.

Risk management supports the appropriate allocation of resources; doing things right the first time is more cost effective. It is also important to understand that risk management is not about removing all risks because this is not possible. It is about understanding the nature of the risks to which the organization is exposed with regard to the achievement of its business objectives, and then implementing sensible cost-effective measures to minimize the downside and (if possible) maximize the upside. When considering risk management through the lens of emergency planning, it is about ensuring the safety of the community that the organization serves and minimizing the impact of any event, incident or disaster should it occur.

3.2 Risk Assessments

Risk assessment is a key component of the risk management process, as shown in Fig. 3.1. A risk assessment can be defined as the process by which information is collected about an event, process or organization in order to identify existing hazards and potential risks.

As well as identifying the consequence (impact) and the likelihood (chance/probability) of the risk being realized, the risk assessment also needs

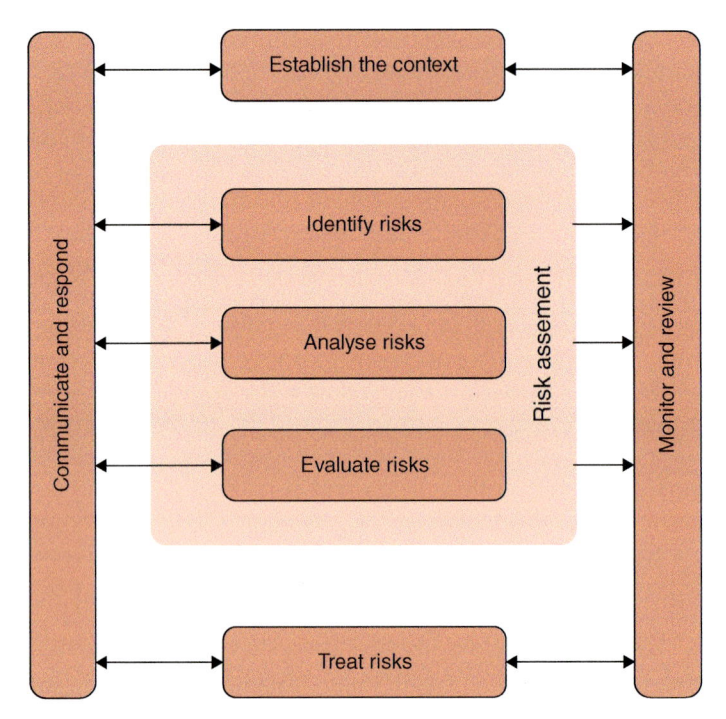

Fig. 3.1. The risk management process.

to consider the control measures that are already in place or that are required to be in place to mitigate or eliminate the risk as far as reasonably practicable. Risk assessment is therefore about identifying and understanding the nature and scale of risks faced by the organization, using three key components:

- *identification* – establish the events, activities, sources and consequences;
- *analysis* – place in a structured format; and
- *evaluation* – rank or rate to provide a measure of the scale of the risk.

The Health and Safety Executive has broken the elements of the risk assessment process down further into five steps, as shown in Fig. 3.2.

Risk assessments can range from simple to complex, and need to be suitable and sufficient for the significance of the perceived risk. It is important to recognize that undertaking a risk assessment is not about creating a huge amount of paperwork and that the level of detail in the risk assessment and the size of the team undertaking the assessment should be proportionate to the risk. When undertaking a risk assessment for an emergency planning event, consideration should be given to the involvement of both internal and external partners and stakeholders.

3.2.1 Identifying potential risks

Key to any risk assessment is accurately identifying the potential hazards or events that may, given circumstances or conditions, present a risk to

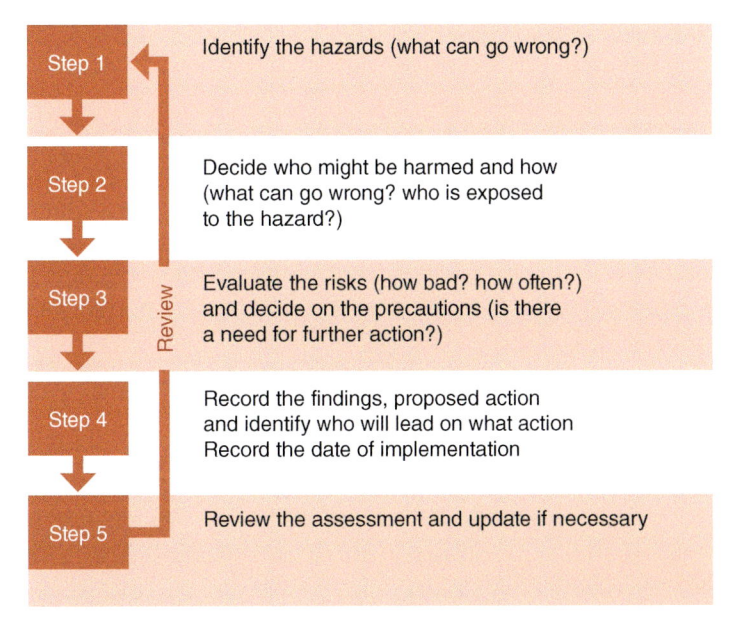

```
Step 1 ← Identify the hazards (what can go wrong?)
  ↓
Step 2   Decide who might be harmed and how
         (what can go wrong? who is exposed
         to the hazard?)
  ↓
Step 3   Evaluate the risks (how bad? how often?)
         and decide on the precautions (is there
         a need for further action?)
  ↓
Step 4   Record the findings, proposed action
         and identify who will lead on what action
         Record the date of implementation
  ↓
Step 5   Review the assessment and update if necessary
```

Fig. 3.2. The Health and Safety Executive's five steps of risk assessment.

the organization. In order to do this successfully, an intimate knowledge of the organization, its organizational objectives and the environment that it operates in is required. In terms of emergency planning, this will also include learning from incidents or events that have happened elsewhere. For an organization to have an effective and methodical approach to risk management and in turn risk assessment, it needs to have a documented, organizational-wide consistent approach.

The information gathered and provided in any risk assessment is critical to the quality of decision making on the allocation of resources to deal with the identified risk(s). This decision making will also be improved if the risks are categorized; example categories are *strategic, operational, legal* and *financial*.

Risks facing any organization can be both external and internal, with risk triggers being proactive or reactive. Figure 3.3 gives some examples of these triggers.

There are a number of techniques available to assist in identifying risk causes, events and effects. From simple to complex, examples of these are:

- brainstorming;
- strengths, weaknesses, opportunities, threats (SWOT) analysis;
- political, economic, social, technology, legislative, environmental (PESTLE) analysis;
- flow charts – looking at each process, describing both the internal processes and external factors that can influence those processes (Fig. 3.4);
- scenario analysis – what ifs;
- root cause analysis – for example, fish bone or the bow tie (Fig. 3.5);
- risk assessment workshops;

- incident investigations;
- auditing and inspection; and
- hazard assessments.

The flow chart process (Fig. 3.4) will help identify events that will disrupt or threaten delivery of a process, perhaps an emergency response. In a flow chart assessment, each stage of the process is identified and consideration is given to the event that may disrupt the stage, what may cause that, the consequences and what control measures are required.

Using the bow tie model (Fig. 3.5) will help identify the causes of an event happening and the effect of the event.

When looking at causes, further analysis can be undertaken by asking 'why' until there is an understanding of why it would happen (i.e. you can't

	Internal	
Internal inspections/audits		Risk assessments
Incidents		Business continuity impact
Complaints		analysis
Claims		Internal stakeholder consultation
Local response to an incident		Internal audits
		Post-exercise reports
National initiatives		Benchmarking
Health & Safety Executive reports		External stakeholder consultation
Safety notices/alerts		National enquiry reports
Public interest reports		Identification of specific threats
Department of Health guidance		Horizon scanning
	External	

(Reactive ← → Proactive)

Fig. 3.3. Possible risk triggers.

(a) Stage 1 → Stage 2 → Stage 3 → Stage 4 → Stage 5 → Stage 6

(b)

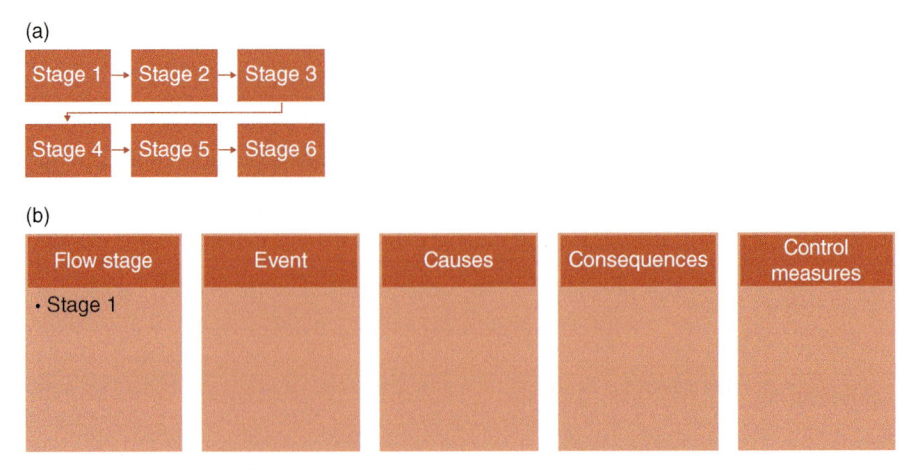

Flow stage	Event	Causes	Consequences	Control measures
• Stage 1				

Fig. 3.4. Examples of flow chart analysis.

Risk Assessment

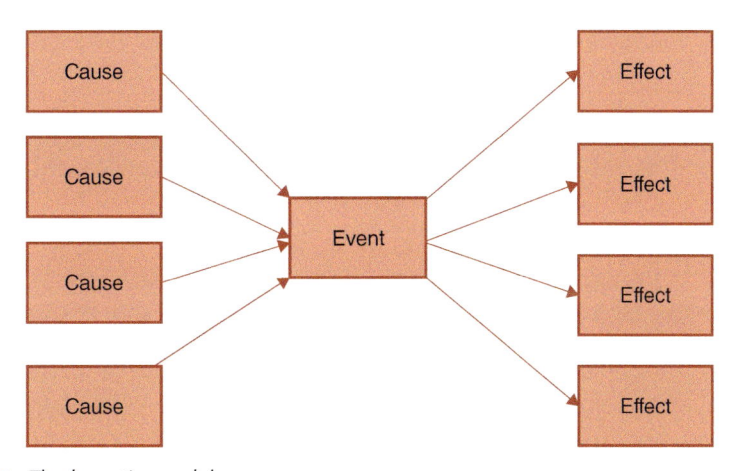

Fig. 3.5. The bow tie model.

answer why any more). Other simple questions to ask when undertaking a risk assessment are:

- What can happen?
- When and where would it happen?
- Who would it affect?
- How many people would it affect?
- How and why would they be affected?

3.2.2 Risk description

Having identified the risk and its possible consequences, the risk needs to be described effectively and clearly. Describing the risk can be a challenge. The objective of the description is that the risk and its impact are clearly understood by the reader, in order to support decision making with regard to further treatment of the risk and allocation of resources. Too high level of description makes it difficult to evaluate the potential consequences (impact) and likelihood (chance/probability) of the risk being realized (e.g. 'staff shortages' is too high level and too vague). Describing the consequences rather than the events or actions that lead to the consequences (e.g. 'failure to meet targets') also needs to be avoided. Therefore, when describing a risk, it needs to be framed in terms of cause and effect.

3.2.3 Evaluating or estimating the risk

It is important to manage risk in a simple and consistent way; therefore, an organizational-wide framework and standard matrix are needed. Earlier in this chapter we talked about risk being measured in terms of consequence (impact) and likelihood (chance/probability):

- *Consequence* is the impact that the risk would have on individuals, the organization and the community should the risk be realized. Consequence of the risk can be described in a spectrum of: *catastrophic, major, moderate, minor* and *negligible/insignificant*.
- *Likelihood* is the probability or chance of the consequence being realized. Likelihood of can be described in a spectrum of: *rare, unlikely, possible, likely* and *almost certain*.

To support the evaluation of the risk a risk matrix should be used, which will enable the overall level of risk to be identified by mapping consequence and likelihood. There are a number of matrices across different industries, ranging from a 3 × 3 matrix (Fig. 3.6) to a slightly more complex 5 × 5 matrix (Fig. 3.7), which provides more a quantitative assessment of both consequences and likelihood.

Consequence	Likelihood		
	Low	Medium	High
High			
Medium			
Low			

Fig. 3.6. A 3 × 3 risk matrix.

	Consequence	Likelihood				
		1	2	3	4	5
		Rare	Unlikely	Possible	Likely	Almost certain
5	Catastrophic	5	10	15	20	25
4	Major	4	8	12	16	20
3	Moderate	3	6	9	12	15
2	Minor	2	4	6	8	10
1	Insignificant	1	2	3	4	5

Fig. 3.7. A 5 × 5 risk matrix.

In order to gain a more comprehensive view of risks, they should be evaluated at three different points of the risk story. These points are:

- *Initial risk score* – the assessment of consequence and likelihood before any controls are in place. This can also be referred to as the *inherent, naked* or *gross risk*.
- *Current risk score* – the assessment of consequence and likelihood with existing controls in place (i.e. the *current risk state*).
- *Target risk score* – the assessment of consequence and likelihood once all the controls (existing and new) are in place and effectively delivering mitigation. This can also be referred to as *net, controlled* or *residual risk*.

Using this concept of three risk scores will show the impact on the risk level of any existing and/or additional controls. It will also help an organization to decide the most effective use of resources. For example, if two risks have the same initial and current score but an investment of £5000 will mitigate one further than the other, the organization can ensure maximum benefit from the financial investment.

Once the risks facing the organization have been evaluated and mapped on a matrix, the organization's work programme can be developed, with the most significant risks that require action or focus being easily identified. However, before decisions can be made regarding how the risk is to be managed, the organization needs to agree its appetite or tolerance for risk; this is known as *risk appetite*. An organization's risk appetite will help identify what the organization is and is not willing to tolerate.

The use of colours and/or numbers in the matrix will also help identify whether the risk can be considered low, moderate, high or critical. This can also support the risk appetite. For example, any risk identified as falling within a yellow or green quadrant may be deemed tolerable, while any risk falling within an amber or red quadrant may be deemed unacceptable.

The risk appetite will help identify what level of actions is to be taken to address the risk. Using the information gathered as part of the risk assessment the organization can agree the most appropriate risk response to managing the risk. These responses can be framed as in Box 3.1.

Box 3.1. Management of risk

- *Tolerate*: accept the risk at its current level.
- *Transfer*: transfer the risk to another party (e.g. by outsourcing or insurance). The consequences of this action will require risk assessing.
- *Terminate*: stop the activity that presents the risk. The consequences of this action will require risk assessing.
- *Treat*: take action to reduce or mitigate the risk, in terms of reducing the likelihood of its occurrence and/or reducing the consequences if it does occur.

When treating, tolerating or transferring a risk, the opportunities that these options present should also be considered. The ranking or scoring of risks will help in identifying the most significant risks, which need to be addressed first. The identification of long-term solutions is important for risks with the biggest consequence; however, it may also be necessary to consider interim actions to mitigate the risk while planning for the longer-term ones.

3.3 Risk Controls

Once the risks have been assessed, the priorities for the organization will emerge, as will the need for development of additional control measures or treatment plans to mitigate the risk further. The objective of any risk treatment plan is to either eliminate or reduce the risk. This can be done by reducing the likelihood or the consequence should the risk be realized.

When considering a risk response, the main response will be to control/mitigate the risk; this can include actions before the risk event is realized (preventive controls) or the development of contingencies or response plans that can be activated should the risk be realized (corrective controls). For example, the risk of major incidents cannot be fully mitigated; however, the clear multi-agency response and command and control structure can help manage any post-incident consequences, ensuring that casualties receive prompt response and reducing the impact of the incident where possible.

When identifying appropriate control measures it is important that an organization does not invest more in any control measure(s) than the cost of the risk it is designed to control. The cost of risk control, therefore, must be balanced against the benefit of controlling the risk.

To ensure wise investment in controls, it is also important to be aware of the effectiveness of the controls. Usually the most effective controls are more expensive to implement than the less effective controls. Effective controls tend to be designed in, are less reliant on people, have backup or contingency arrangements, are clear and easy to comply with, and are well communicated and understood, supported by good training and by management. Once the controls have been identified agreement on who is able to deliver or influence delivery of the required actions needs to be reached.

It is also important that assurance mechanisms are identified that will provide assurance that the controls are effective and delivering the mitigation expected. Assurance can be positive, showing that the controls are working, or negative, showing that they are not working. Assurance can be achieved in a number of ways including:

- audit;
- scenario testing or exercises (Chapter 11, this volume); and
- development of key performance indicators, such as number of incidents and/or complaints.

3.4 Risk Reporting

The outcome of risk assessments is the development of a risk register. A risk register is merely a repository of risk information; it is a dynamic tool for monitoring the actions being taken to mitigate the risks as well as recording the controls already in place.

A risk register will provide quantified and ranked risks, thereby giving the risk profile of the organization. It informs the direction of managing risks and tracks delivery of the mitigating actions. It is also a tool that can be used to communicate the risks faced by the organization or the community. An emergency preparedness risk register will help identify the key vulnerabilities of an organization or a community. It will also help identify risk themes and so inform the type of emergency plan that needs to be developed. An example risk register template is presented in Fig. 3.8.

The risk register can be then used to prioritize and develop the organization's work plan, ensuring that the right areas are focused on.

3.5 Building Preparedness

Once the organization has undertaken its risk assessments, it will understand which risks are most likely to occur within the community in which it operates and which risks are likely to create the biggest impact, thus enabling the development of a work plan.

The resulting work plan will then provide a route to prioritize the preparedness arrangements for the highest risks but also give due time and effort to the lower risks. The balance is between ensuring that, as a priority, there are arrangements for the high-risk events while ensuring that the lower but no less important risks are addressed within an acceptable time period.

The risk assessment process is designed to provide a mechanism to understand the context to emergency planning. Where risks can be migrated they should be and where the risk cannot be eliminated there should be plans in place to safely and effectively respond to any resulting emergency.

Key Answers

- Risk can be described as the uncertainty of outcome, be it positive (an opportunity) or negative (a threat), if an event or emergency occurs.
- The relevance or priority of a risk can be calculated by an assessment of the likelihood of an emergency occurring against the impact if it does occur. Take, for example, the medical risk assessment for a public air show. The impact of a plane crashing into the public is significant, against the likelihood being very low. However, the bouncy castle at the same event, which has a relatively high likelihood of some grazes and bruises, has a very low impact. Therefore, both examples are catered for in the event's medical plan.
- Once a risk has been assessed and quantified, a relative score can be applied that helps to prioritize the attention given to it. Work can then be focused on dealing with the high risks first, by mitigating the risk by applying specific measures, planning for effective response arrangements or a combination of both.
- Once the risk of an emergency occurring has been assessed, the risk rating is applied. By plotting each of these risk ratings together in a single table, the emergency planner can review this against his/her organization's capability to respond. If the capability to respond to a high-risk emergency is low this will inform the priority that the emergency planner needs to give to building capability. Lower-risk emergencies can be planned for in slightly slower time so that high-risk events can be given priority.

Risk ID	Owner/ lead	Risk description	Risk category	Initial score	Existing controls	Current score	Risk response	Additional controls			Target score	Review date
								Control	Progress	Due date		

Fig. 3.8. Risk register template.

Further Reading

Cabinet Office (2011) Keeping the country running: natural hazards and infrastructure. Available at: www.gov.uk/government/publications/keeping-the-country-running-natural-hazards-and-infrastructure (accessed 31 December 2015).

Cabinet Office (2012) Emergency preparedness: Guidance on part 1 of the Civil Contingencies Act 2004, its associated regulations and non-statutory arrangements. Chapter 4. Local responder risk assessment duty. Available at: www.gov.uk/government/publications/emergency-preparedness (accessed 31 December 2015).

Cabinet Office (2013) Risk assessment: how the risk of emergencies in the UK is assessed. Available at: www.gov.uk/guidance/risk-assessment-how-the-risk-of-emergencies-in-the-uk-is-assessed (accessed 31 December 2015).

HM Treasury (2004) The Orange Book. Management of Risk – Principles and Concepts. Available at: www.gov.uk/government/publications/orange-book (accessed 28 February 2016).

Hopkin, P. (2014) *Fundamentals of Risk Management*, 3rd edn. Kogan Page Limited, London.

Joint Technical Committee OB-007, Risk Management (2013) *Risk Management Guidelines – Companion to AS/NZS ISO 31000:2009, SA/SNZ HB 436:2013*. Standards Australia/Standards New Zealand, Sydney, Australia/Wellington, New Zealand.

4 Writing an Emergency Plan

A. WAPLING

Regional Head of Emergency Preparedness, Resilience and Response, NHS England (South), UK

Key Questions

- Why do we need an emergency plan?
- Why can't we just use existing processes during an emergency?
- What process should be adopted to write a good plan?
- How will we know if the plan is effective and fit for purpose?

4.1 Why Do We Need Plans?

Surely in an emergency we just do more of what we normally do? We simply apply the usual organizational processes to manage an incident, don't we?

By their nature, emergencies are not very frequent, often occur without warning or notice, and can be extreme. Therefore, our response to emergencies needs to have been considered outside the normal operation of the organization because the response to these emergencies will be unfamiliar to most and require arrangements that are over and above what the organization traditionally delivers.

An ambulance service (emergency medical service) will have day-to-day arrangements for managing the response to 999 (911) calls by the dispatch of appropriate clinical resources. On the whole, these responses deal with a small number of sick or injured people. This is a regular day-to-day operation for all staff involved. There is, however, a need to ensure that there are arrangements in place for those less frequent emergencies where an ambulance service needs to respond to more injured people than it has the direct resources to cope with. An example of this would be the London bombings in 2005 and the terrorist attack in Mumbai in 2008.

For these extreme events, responding organizations need a predetermined framework to ensure that those survivors get the best possible and most appropriate care in such challenging circumstances, but also that the responders have an agreed framework to respond with to support their decision making and to give them the boundaries to cope with such a traumatic event emotionally.

The predetermined framework to help both those caught up in the emergency and the responders is the emergency plan. Time spent working through a scenario with partner organizations ahead of any emergency is time well spent. The more planning that can be carried out in advance, the better the response will be for both the survivors and the responders. Therefore, the purpose of an emergency plan is to provide a predetermined framework for responding organizations to deliver a safe and effective response to an emergency in order to maximize the positive effect on those involved.

There is, however, a fine line between an overly detailed and prescriptive plan and a plan that is so vague it provides no benefit. An overly complex and wordy plan is unlikely to be read in advance and even less likely to be read in an emergency when time is precious and anxiety is high. A complex and wordy plan is often understood only by the person who wrote it. An overly detailed and prescriptive plan will often constrain the emergency responders in their ability to make responsive and creative decisions in reaction to an evolving response.

Responders need the safety of a pre-agreed framework to make safe and creative decisions within the response. No emergency ever plays out in the same way as we believe it will in the planning stages, therefore the plan needs to provide the flexibility to be adapted to cater for the presenting scenario. Too prescriptive and the plan will not be picked up and used because the presenting emergency doesn't exactly fit that described within. Equally, a plan that leaves too much room for reactive determination will not provide the boundaries that the responders require nor the triggers required to effect the right reactions at the right time for the best outcomes for all those involved.

Therefore, time should be given to the level of detail contained within a plan before committing words to the page. This can often be done by engaging with those who will be the recipients of the plan, both in terms of the responding staff as well as members of the public.

An emergency plan should therefore follow a core set of principles:

- It should be appropriate for the intended audience, including the appropriate use of language and terminology so that there is no ambiguity or confusion.
- It should be current, up to date and relevant to the organization and its structures (including that of partner responding organizations). Our interpretation of risk changes, science and our understanding of incidents

progress, and the learning from incidents and exercises informs our arrangements; therefore, plans must always be current.
- It should have appropriate governance arrangements for the authority of the plan. Who has signed the plan off to ensure that it has gone through the appropriate rigour and challenge? Has the organization formally agreed that this is the framework it will take in an emergency?
- It should set out responsibilities for the plan and how the plan works. When a plan needs to include some detailed and predetermined actions, they need to be presented in a way that can easily be found and followed. The beginning of the plan should include some mechanism for the user to find the relevant details when he/she is under pressure and working in a challenging environment. The plan needs to set out its responsibilities in terms of what is in scope and what is not. What emergencies will the plan be relevant for and which ones not (because they are covered elsewhere)?
- It should link to local, regional and national risk registers so that the user of the plan can understand the context.
- It should link with other relevant plans, including that of other agencies. How does this particular plan sit with plans such as the organization's internal business continuity plan, especially when responding to an external incident?

There are in principle three types of emergency plan: (i) the generic or core plan; (ii) the threat-specific plan; and (iii) the business continuity plan. All three types of plan should be written with the other in mind. Although they may be stand-alone plans, there does need to be an overt link to and acknowledgement of the other. This will support users in navigating what can end up as a complex suite of plans. This chapter describes the generic/core and threat-specific plans; Chapter 10 contains details on business continuity plans.

4.2 Generic/Core Plans

The generic/core plan is the overarching plan that provides the baseline response to all emergencies. It contains the management processes that any threat-specific emergency can be managed within. These generic/core plans can be single-agency or multi-agency, either setting out the single organization's management processes or that of a partnership multi-agency response. A generic/core plan should contain the following elements:

- A clear statement of intent. What is the authority of the plan and the remit of the agency/agencies committed within this plan? What is the legal framework within which the plan operates?
- A defined mechanism for enacting or activating the plan. Who can activate the plan? How do they activate the plan? What is the route for escalation? What are the triggers that could be applied to aid decision making to activate the plan?

- Clearly defined roles and responsibilities for those identified within the plan. Who is responsible for what? And what is their scope for decision making and actions?
- A specified management process for authority of decision making, also known as a command structure (as described in Chapter 7). Who is responsible for which level of decisions, so that there is no ambiguity or confusion?
- A process for collecting information in order to inform decision making and enacting the response to decisions, also known as control (as described in Chapter 7). What are the management processes needed to collect relevant information? Does this require the establishment of an incident coordination centre or a control facility? What are the management processes needed to operate this facility?
- A process for sharing information within the organization, across to other organizations or with the public, also known as communication (as described in Chapter 8). How is information shared and with whom? What are the governance arrangements for sign-off of information prior to circulating? Who are you communicating with and how (verbal, e-mails, briefings, websites, etc.)? How are you managing your relationship with the public?
- A framework for working with and operating as a collective with partner organizations, also known as coordination (as described in Chapter 7). What are the protocols for working with key responding organizations? How are those relationships managed and maintained in an emergency?
- Triggers and actions that can be used for de-escalation and stand down from the response elements of an emergency. Who is responsible for the de-escalation from an emergency and what are the steps that need to be taken to manage this process?
- A process for managing the recovery from an emergency. Not only the recovery of a community or of patients but also of the responding organization, because the disruption from an emergency for a responding organization can be significant (see Chapter 12).
- A policy for managing and preserving information following an emergency. How is information collected, collated and stored safely for the required amount of time should information be required for any legal processes that may follow?
- A framework for identifying the lessons that have been generated from the response to an incident. What is the debriefing process? What process should occur by when? And by whom? When should a report be published? How will lessons be learned and incorporated back into planning?
- A process to support staff following an emergency in terms not only of psychosocial support, but also for any possible legal processes that may follow.

4.3 Threat-specific Plans

Threat-specific plans are often required to provide a specific layer of detail that is unique to the type of incident. An example would be that of a radiation incident, which would need specific information in relation to the hazard, the treatment of those exposed and the long-term consequences. A threat-specific plan should be written in the context of the generic/core plan but provide the unique context and actions framework required to respond to this type of emergency. A threat-specific plan may also be written for a specific building or site, for example a concert venue, sports ground or large industrial chemical plant. A threat-specific plan may contain the elements listed below and be laid out as in Box 4.1:

- a more explicit description of the risk and the science behind it;
- more detail on the possible consequences of the threat;
- explicit actions and options that are available to responding organizations to effect a safe and effective response;
- a clear route to gain expert advice and support to aid decision making and response activities; and
- the roles and responsibilities of agencies that are unique to the response to the specific threat.

Box 4.1. Suggested plan structure

- Aim and objectives of the plan.
- Trigger for activation and activation procedures.
- Identification of roles and responsibilities for all parts of the organization.
- Location of incident coordination centre.
- Roles and responsibilities of other responders.
- Trigger for deactivation and deactivation procedures.
- Stand-down procedures.
- Training and exercise schedule.
- Information about the specific hazard or site.
- Identification and roles of staff and/or multi-agency teams.
- Identification of lead responsibilities of different responder organizations.
- Location of joint operations centre (if applicable).
- Communications procedures or plans.
- Contact details of key personnel and partner agencies.

4.4 Stakeholders

No plan can be written in isolation nor should it be done without consultation. The plan needs to reflect the environment the response will be made within. The response to a large road traffic collision resulting in a significant number of casualties cannot be made without the police, fire and rescue services, hospitals, ambulance service, highways agencies, local government organizations, etc. Therefore, no plan designed to respond to such an incident can be written without engaging the relevant stakeholders and plan users. Time should be given in advance of writing the plan to undertake meaningful discussions with all other responding organizations so that a mutual picture can be understood as to who owns which responsibilities and actions. A lack of that picture can result in organizations believing that the other is responsible for undertaking an action and therefore the action is not delivered – resulting in deterioration of the situation or indeed the loss of life. Alternatively, not knowing another agency is going to undertake an action can result in a number of organizations delivering this action. This can result in wasting of resources, loss of valuable time and general confusion, as well as loss of public confidence.

The author has reflected on the planning in advance of the 2009/10 influenza pandemic, where much time was taken with partner organizations talking through the various scenarios and painstakingly identifying which agency was responsible for the delivery of which intervention. The analogy was made to walking down a corridor of doors and when you opened each door behind it was another corridor full of doors. Much time was taken opening each of the doors, engaging with partners to understand what was found behind that door and articulating the actions that would then need to be defined.

4.5 Identifying Gaps in the Plan

No plan should be written and published without some form of challenge and testing, nor should it be left on the shelf for a number of years without review or updating.

There are a number of ways in which a plan can be challenged to identify if there are any gaps or omissions. These can include:

- sharing your plan with peers for review;
- comparing your plan with others;
- sharing it with users of the plan;
- formal exercising via a tabletop or live exercise (described in Chapter 11);
- undertaking a scenario-based discussion;
- reviewing the response to incidents and emergencies; and
- reviewing the response to exercises of other agencies.

4.6 Summary

The plans we have must be effective and serve those whom we are responding for and those who will be effecting the response. A plan for the planner's sake is an ineffective plan.

> **Key Answers**
> - Plans are needed to provide a safe framework to ensure an effective response.
> - Many of our local day-to-day process would not stand the pressure of an emergency response and during the stress of a response is not the time to invent new systems.
> - This chapter outlines a framework to develop plans. No one process will fit all plans; however, the chapter has provided a starting point that can be adapted as necessary.
> - Exercising plans helps to identify gaps in planning and determine if the plan is effective and fit for purpose.

Further Reading

Cabinet Office (2006) Emergency preparedness. Guidance on part 1 of the Civil Contingencies Act 2004, its associated regulations and non-statutory arrangements. Chapter 5: Emergency Planning (revised October 2011). Available at: www.gov.uk/government/publications/emergency-preparedness (accessed 18 December 2015).

Canadian Centre for Occupational Health and Safety (2014) Emergency Planning. Available at: www.ccohs.ca/oshanswers/hsprograms/planning.html (accessed 10 December 2015).

Department of Homeland Security (2016) Emergency Response Plan. Available at: www.ready.gov/business/implementation/emergency (accessed 18 December 2015).

Health and Safety Executive (2016) Emergency procedures. Available: at www.hse.gov.uk/toolbox/managing/emergency.htm (accessed 10 December 2015).

5 Emergency Planning and Response: Working in Partnership

S. Lewis

Head of Crisis Response, British Red Cross, London, UK

Key Questions

- Why is it important to plan in partnership?
- What are the benefits to responding organizations of working in partnership?
- What are the benefits to people caught up in an emergency if organizations work in partnership?
- Which are the key partner organizations for health?

5.1 Introduction

Working in partnership when planning for, responding to and recovering from emergencies is not only recognized as best practice, it is now seen in many quarters as business as usual. While the quality of the partnerships may vary (some are formal and some less so), all are based on relationships, and it is these relationships, built up through planning, training and exercising, which pay dividends during the response to a major incident or emergency. This can all be summed up in one word: trust. Trust in an individual, trust in an organization and trust in a tried-and-tested team approach.

Working in partnership while planning to respond to major incidents allows each agency to fully understand one another's remit and, through this, to identify any duplication or gaps that could impact on the response to people caught up in an emergency. While all agencies have well-rehearsed individual plans and procedures, there are real benefits to planning, training and exercising as a multi-agency partnership. Planning for and responding to emergencies is one area where public, private and voluntary sector organizations are often able to put their different cultures aside for the very real benefit of meeting the needs of people in an emergency.

This chapter expands on the personal experiences of the author over many years in emergency preparedness and response working in two very different organizations (the Metropolitan Police Service in London and the British Red Cross). Both regularly form partnerships with the health services to good effect. The specific examples in this chapter are based on personal experience; however, the principles can easily be extrapolated to any other combination of partnerships.

5.2   The Civil Contingencies Act 2004

In 2004, the UK Civil Contingencies Act was passed through parliament. The Act identifies various statutory multi-agency partners as Category 1 and Category 2 responders (Table 5.1) and identifies different roles and responsibilities attributable to them (Table 5.2). Additionally, it gives their chief officers vicarious responsibility to ensure that they plan, prepare, train and exercise for emergencies, individually and in partnership.

In addition to those described above, there are many other organizations and groups that are involved in planning for and responding to major incidents. For example, the voluntary sector is not included in either category; however, the Act states that Category 1 and 2 responders must give regard to

Table 5.1. Category 1 and Category 2 responding organizations.

Category 1 responders	Category 2 responders
Emergency services	Utilities
Police forces	Electricity distributors and transmitters
British Transport Police	Gas distributors
Fire authorities	Water and sewerage undertakers
Ambulance services	Telephone service providers (fixed and mobile)
Maritime and Coastguard Agency	Transport
Local authorities	Network Rail
All principal local authorities (i.e. metropolitan districts, shire counties, shire districts, shire unitaries)	Train operating companies (passenger and freight)
	London Underground
Port Health Authorities	Transport for London
Health bodies	Airport operators
Acute Hospital Trusts	Harbour authorities
Foundation Trusts with emergency departments	Highways Agency
	Government agencies
NHS England	Health and Safety Executive
Local Health Boards (in Wales)	
Any Welsh NHS Hospital Trust that provides public health services	
Public Health England	
Government agencies	
Environment Agency	

the capacity and capability of the voluntary sector. Future reviews of the Act may give consideration as to whether greater emphasis should be placed on more formalized arrangements between statutory responders and other partners that are not identified specifically in the current guidance. Case Study 5.1 is a good example of partnership working within the London Resilience Team, a precursor to Local Resilience Forums.

5.3 Local Resilience Forums

The Civil Contingencies Act 2004 identifies the need for formal structures to enable multi-agency partners to engage and plan with each other. Following the successful establishment of the London Resilience Team in 2002, Local

Table 5.2. Roles and responsibilities of Category 1 and Category 2 responders.

Category 1 responders	Category 2 responders
Risk assessment	Cooperation
Business continuity management (BCM)	Information sharing
Emergency planning	
Maintaining public awareness and arrangements to warn, inform and advise the public	
Provision of advice and assistance to the commercial sector and voluntary organizations (local authorities only)	
Cooperation	
Information sharing	

Case Study 5.1. The London Resilience Team

In 2002, following the terrorist attacks in the USA on 11 September 2001, the London Resilience Team (LRT) was formed. This consisted of senior civil servants seconded from several government departments, together with experienced practitioners and senior representatives seconded from responding agencies including police, fire, ambulance, local authorities, the military, the NHS, the Health Protection Agency (now Public Health England), the voluntary sector, business representatives, transport operators, the Port of London Authority, the Maritime and Coastguard Agency, and utility companies.

The team was governed by a formal meeting structure and membership, and its primary role was the development of multi-agency emergency plans to respond to a major incident or emergency. These plans included: Mass Fatality, Mass Casualty, Mass Evacuation, Warning and Informing, CBRN (chemical, biological, radiation and nuclear), and Site Clearance.

Numerous benefits were realized through this multidisciplinary team coming together to develop plans, train, exercise and respond. This included the delivery of coordinated multi-agency responses across London during periods of severe weather (such as heavy snowfall across London in January 2013 and localized flooding in Croydon in south London in February 2015) and during the 2009/10 influenza pandemic.

Resilience Forums (LRFs) have been established across the UK. The London LRF was established following publication of the Civil Contingencies Act 2004 and has met regularly to discuss preparedness as well as respond to many incidents and pre-planned events in London.

LRFs in England and Wales are based on police service area boundaries and all have largely consistent membership and work plans, which are tailored to reflect local organizations and priorities. LRF equivalents are also in place in Scotland and Northern Ireland.

A key role of the LRF is the oversight of strategic emergency preparedness for the LRF area that reflects the roles and responsibilities of all partners. Often underpinning the LRFs are themed panels (such as Blue Light, Utilities, Business and Voluntary Sector), the Chairs of which attend the main LRF to represent their group. There can also be a number of Task and Finish Groups that undertake the development or review of strategic emergency plans for the LRF area and will include the most appropriate membership from the panel groups above.

The LRF Chair is typically the Chief Executive or Chief Officer of a Category 1 responder, most often police or fire. As with any process that is replicated many times, there are differences between the LRFs. Different agency chairs and different personalities come to the fore, area risks help focus work plans and can help forge better partnerships. For this to work well there needs to be good relationships across and between LRFs.

5.4 Local Health Resilience Partnerships

In April 2013 the Health and Social Care Act strengthened arrangements for health emergency preparedness, resilience and response (EPRR) through the establishment of Local Health Resilience Partnerships (LHRPs). This meant that at LRF level, the coordination of health EPRR was aligned with multi-sectoral emergency preparedness and response and mirrored the LRF boundaries. The Chair or a representative would also attend the LRF. LHRPs are co-chaired by the local Director-level Accountable Emergency Officer (AEO) from NHS England and the local Director of Public Health (DPH). Membership of the LHRP includes providers of NHS-funded care (including acute and tertiary hospitals, mental health, community providers and ambulance trusts; and where private and voluntary sector organizations are delivering NHS-funded care, they are expected to attend), health commissioners, local authority public health departments and Public Health England (PHE).

The principal roles and responsibilities of the LHRP, as set out in guidance from the Department of Health, are described as:

- Provide a strategic forum for local organizations (including private and voluntary sector where appropriate) to facilitate health sector preparedness and planning for emergencies at LRF level.
- Facilitate the production of local sector-wide health plans to respond to emergencies and contribute to multi-agency emergency planning.
- Be coterminous with LRFs.

- Provide support to the NHS, PHE and local authority representatives of the LRF in their role to represent health sector EPRR matters.
- Provide support to NHS England and PHE in assessing and assuring the ability of the health sector to respond in partnership to emergencies at an LRF level, and may provide similar support to local authorities and DPH where appropriate.
- Each constituent organization remains responsible and accountable for the effective response to emergencies. Similar to LRFs, the LHRP has a pivotal role in facilitating planning but does not have a collective role in the delivery of emergency response.

5.5 Partnerships within the Health Sector

The health sector is comprised of a wide variety of organizations, including commissioners and providers from across the public, private and voluntary sectors. Health organizations provide a variety of services to patients including primary care, secondary and acute care, specialist tertiary care, physical and mental health care. Care is provided in the community as well as in health care premises, and at all hours of day and night.

It is essential that organizations build robust tried-and-tested local relationships in order to be able to respond to any major incident or emergency where the population requires access to health care. Some organizations may cover small geographic locations and have a largely pre-identified patient base (such as a primary care general practice), whereas others cover much larger areas and serve a fluctuating population (such as ambulance services) and may not have any prior information on their patients.

It is important that not only do health organizations understand their partner health organizations, but also that multi-agency partners also understand which organizations provide which services. Case Study 5.2 provides an excellent example of multi-agency working and learning following the response to the terrorist attacks in London in July 2005.

Case Study 5.2. The 7/7 bombings, London

The terrorist attacks in London on 7 July 2005 are a key event in the evolution of multi-agency response in the UK. To this day, lessons identified from these experiences continue to inform ongoing preparedness and response to major incidents and emergencies in London and further afield.

The terrorist attacks comprised four separate incendiary devices on the London bus and underground rail transport network during the morning rush hour. Fifty-two innocent people lost their lives and over 700 people were injured in this attack. The four scenes were complicated by the fact that they straddled different local authority areas (Westminster, Camden and the City of London, Fig. 5.1) and police services (Metropolitan Police Service, British Transport Police and City of London Police).

Continued

Case Study 5.2. Continued

Within 90 min of the incident, the first multi-agency Strategic Coordinating Group meeting was held. There were over 20 people present, each representing different organizations with key roles in the response including police, fire, ambulance and the affected local authorities.

The initial response, to ensure the sites were safe and to care for and retrieve patients for transport to hospital, was completed relatively quickly.

However, many of the responding organizations remained involved for many days, weeks and months through the process of collecting evidence, restoring normality, reassuring the public, and providing care and treatment to those injured or affected in the incidents.

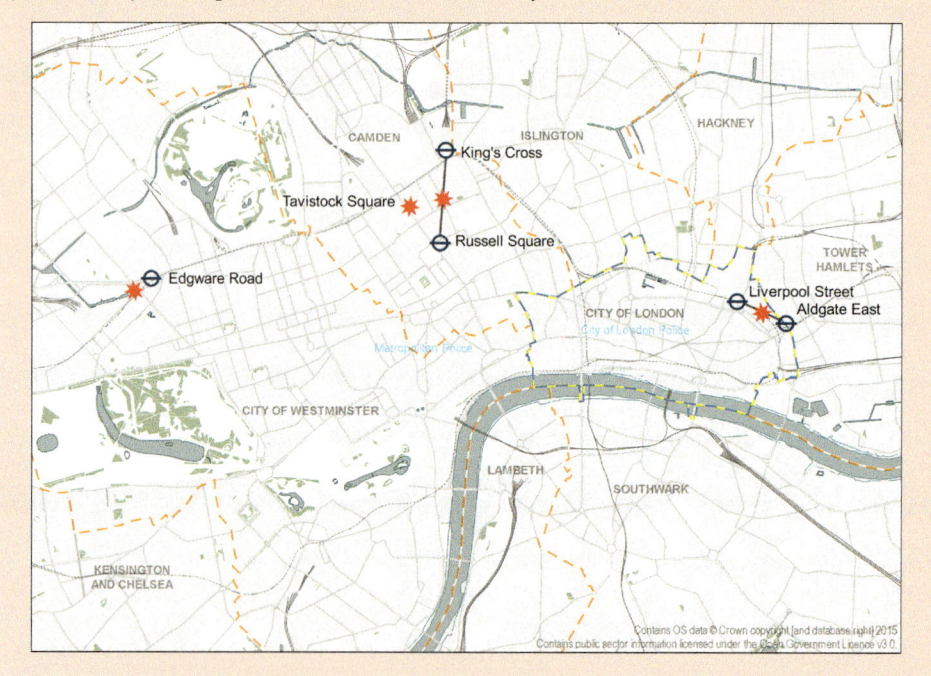

Fig. 5.1. Locations of the 7/7 incidents.

5.6 Importance of Geography, Public Perception, and Capacity and Capabilities

In a major incident, most people will not differentiate between several providers of the same service, and this emphasizes the importance of ensuring all have the same information and overall aims. During the internal briefing for a large event, one senior police officer reminded his audience that the person receiving a service from the police will not look at cap badges to identify which particular service they are from; the person will just expect a professional service from a police officer. This also applies to providers of NHS services, where, in a major incident, members of the public may just see the NHS logo and seek care.

Understanding agency remits and how they can work best together is a key reason to work in true partnership; however, with all manner of major incidents regularly failing to conform to agency geographical boundaries, this is even more vital. While partnership working across different agencies is one aspect, just as important is partnership working within agencies across borders.

Organizational artificial boundaries, geography and jurisdiction are not the only reasons why working in partnership is essential. No one organization has the capacity and capability needed to deal with every aspect of the response to or recovery from an emergency, and planning together in advance can identify any gaps or duplications. Both are equally important to resolve in advance rather than during a response, when lives may be put unnecessarily at risk if there is a lack of clarity.

5.7 Building Relationships

Good partnerships can be formed by building professional relationships in a variety of ways: training and exercising; informal and formal meetings; and planning and responding together. Getting to know key partners in a face-to-face interaction is invaluable. The trust that is built in this way between people and agencies comes into its own during the response phase.

Referring back to the 7/7 London bombings, during the previously mentioned Strategic Coordinating Group meeting and while there was still a genuine fear that further attacks were imminent, the chief police officer chairing the meeting quietly reflected how reassured he felt given he knew everyone around the table by first name and that trust was already there through previous partnership working, responses, training and exercising. Building this level of trust relies on personalities, passion and commitment, and it is to be supported and encouraged wherever possible. A perfect example of organizations learning to improve how they work together is the development of the London Emergency Services Liaison Panel, as described in Case Study 5.3.

5.8 The Importance of the Voluntary Sector

In thinking about multi-agency partnerships when responding to an emergency, it is important to remember that the voluntary sector itself is complex and composed of many different local, regional, national and international charities with varied levels of emergency planning training, experience, equipment, capacity and capability. The sector has an extremely diverse remit and a range of skills, experience, governance and coverage. Some are specifically health focused, while others more humanitarian or faith based; however, all play key roles in planning for and responding to emergencies.

Ideally, all appropriate voluntary sector organizations would be represented at each LRF; however, not all voluntary sector organizations have the capacity to do so. Therefore, voluntary sector organizations typically

> **Caes Study 5.3.** London Emergency Services Liaison Panel
>
> The London Emergency Services Liaison Panel (LESLP) came into existence in 1973 through the lessons identified from blue light agency responses to Irish Republican Army (IRA) attacks on the UK mainland. While police, fire and ambulance services responded well to the same incidents, they did so disparately.
>
> The Panel drew up a guidance manual which the police, fire and ambulance services in London signed up to. The guidance focused on ensuring absolute clarity about agency roles and responsibilities, including command and control. Over time, the Panel has grown in recognition of the multi-agency support and engagement necessary in meeting the needs in any humanitarian emergency. LESLP membership now includes the Metropolitan Police Service, the City of London Police, representatives of the British Transport Police, the London Fire Brigade, the London Ambulance Service, local authorities, the Port of London Authority, the Maritime and Coastguard Agency, the military and the voluntary sectors.
>
> The LESLP manual is regularly reviewed and revised, and consequently is recognized by many as best practice. The recently devised Joint Emergency Services Interoperability Programme (JESIP) is a true sign of how vital partnership collaboration is and how it is necessary to formalize structures. In many ways this could be seen as the first step towards nationalizing the LESLP.

engage with LRFs via a voluntary sector panel as described earlier. This is most effective where the nominated Chair represents the whole group on the main LRF and is strongest where a formal agreement is in place and all organizations sign up to a minimum level of competencies so that partner agencies know what to expect.

In the UK, the national Voluntary Sector Civil Protection Forum (VSCPF) was convened nationally in 2001 by the British Red Cross (BRC) at the request of the Cabinet Office. This was designed to help promote the capacity and capability of the voluntary sector and to provide advice on this in peace time and during a response to an emergency. Members include the Cabinet Office, the Department of Communities and Local Government (DCLG), the Local Government Association and the National Police Chiefs' Council (NPCC). Voluntary sector members include BRC, St John Ambulance, the Salvation Army, Cruse Bereavement Care, Victim Support, Radio Amateurs' Emergency Network (RAYNET), the Royal Voluntary Service, 4 × 4 Response UK, and Search and Rescue England and Wales (Fig. 5.2). The group is constantly looking to improve how the voluntary sector works together and how it can do more with the statutory partners in supporting people in crisis.

5.9 The International Red Cross and Red Crescent and the British Red Cross

As the world's largest humanitarian network, the International Red Cross and Red Crescent Movement provides a long-standing example of interoperability, systematically connecting the capabilities of local volunteers

Fig. 5.2. Working Group Members of the Voluntary Sector Civil Protection Forum.

and international responders to meet the needs of affected communities. The Movement encompasses the International Committee of the Red Cross (ICRC), the International Federation of Red Cross and Red Crescent Societies (IFRC) and national societies in 189 countries. Each national society also serves as officially recognized auxiliaries to their governments in the humanitarian field.

While interoperability within the Movement has evolved over the last 150 years, it remains as relevant as ever. This has most recently been demonstrated through the response to Ebola in West Africa where the BRC and partners worked together, as well as with governments, the private sector, the military, the NHS and other voluntary sector organizations such as Save the Children, to enhance the effectiveness of the collaborative response to the needs of affected people.

The good partnerships in emergency planning described above are absolutely integral to the work of the Red Cross organizations. In the UK the BRC route to people who need support is always with or for the UK statutory services. Without good partnerships and relationships, it would not be possible to meet this remit. Partnerships within the voluntary sector are vital as all bring a wealth of different skills and experience that enhance, complement and support each other (Fig. 5.3).

Fig. 5.3. Partnership working in action.

At the Humanitarian Assistance Centre set up after the 7/7 terrorist attack on London in 2005, the BRC, Salvation Army, Victim Support and Cruse Bereavement Care worked together with Westminster City Council to support people in crisis. This is a perfect example of partnership working at a single centre providing victims and bereaved families with a comfortable place to access humanitarian support from a wide range of voluntary and statutory providers.

5.10 Corporate Partnerships

It is not just statutory and voluntary sector organizations that work together to prepare for and respond to major incidents. Many corporate and private sector organizations are also active in this field; for example, the BRC has a number of corporate partnerships including operational arrangements with a 4 × 4 vehicle manufacturer and a major supermarket.

In the UK, the voluntary organizations regularly respond to emergencies related to extreme weather events, such as flooding caused by heavy rainfall, ice or snow. This is especially so in rural settings where many roads are cut off, leaving individuals and communities vulnerable or more vulnerable. In these situations, four-wheel drive vehicles are essential and while voluntary organizations have several such vehicles, it is not always possible to ensure that the sufficient number is in the right area of the UK at the right time to respond to all requirements. The arrangement with a 4 × 4 vehicle manufacturer is such that in an emergency the BRC can access four-wheel drive hire cars at any dealership in the UK free of charge.

Similarly, the arrangement with a major supermarket is also particularly useful. As a major UK supermarket it is present in many towns and cities, and many stores are open 24 hours per day. In an emergency the BRC can obtain goods to the value of £500 from any store at no cost. Clearly, different emergencies have different needs, and access to items such as food, water, torches, batteries and blankets means the BRC is able to support people in crisis quickly and with lower requirement for the BRC to hold large quantities of stock.

5.11 Government and Statutory Agency Partnerships

Following the Asian tsunami on 26 December 2004, the UK Foreign and Commonwealth Office (FCO) formed a relationship with the BRC to provide psychosocial support to bereaved families and victims of crisis abroad. Since this incident, an on-call team of trained volunteers, many of whom are clinical psychologists, remains on standby to deploy with the FCO Rapid Deployment Team in a future emergency where their support is needed. This is a perfect example of the auxiliary to government role mentioned earlier.

In a UK emergency involving a large number of seriously injured people or fatalities, or in a similar incident outside the UK but involving UK nationals, a Casualty Bureau is opened by the police service. This is a part of the major investigation and is in place primarily to trace and identify people involved in an incident and to reconcile missing person records with casualty and survivor/evacuee records. The BRC has a Support Line capability and a memorandum of understanding with the NPCC so that it can be launched to complement the NPCC Casualty Bureau. The Support Line has the capacity and capability to be fully scalable and includes a bank of trained volunteers together with a 24/7 telephony system able to route callers from the Casualty Bureau or directly to the Support Line.

5.12 Summary

This chapter uses real-life examples of multi-agency partnerships and incidents to illustrate the benefits of working in this way. The formation of some of these formal partnerships and relationships has been as a result of learning from major emergencies and has genuinely benefited subsequent responses and in turn those people affected.

Making and building relationships and partnerships is not always easy. Having and focusing on a common goal is key; however, for some individuals or organizations the goal may not seem as apparent or the need for a partnership as relevant as to others. It probably seems obvious that blue light agencies such as police, fire and ambulance services work with one another, and yet examples such as the major supermarket's arrangement with the BRC are less obvious. Some partnerships evolve due to the common goal of meeting the humanitarian need of people in a crisis.

Often it will take a joint response to an incident for individuals and organizations to see the real benefit. It is always important to consider barriers to entry and what might prevent an organization being part of a response to an emergency. If another provider with similar capacity and capability is in that space and people's needs are being met, then that is less concerning. If, however, it is because an organization has been forgotten or excluded then remedial activity must be undertaken.

Working in true partnership makes perfect sense in terms of an efficient and effective response to an emergency. It is essential that focus is maintained on the people caught up in an emergency, and that we recognize and remember that working in true partnership with appropriate agencies affords the best solution in the planning, response and recovery from emergencies for all involved.

Key Answers

- If emergency responders do not plan in partnership, the relationship and trust will not be in place for the response. Equally, as the response is a true multi-agency partnership (as no one responder has the remit to cover all aspects of a response), the same has to be true of the planning phase.
- The benefits to responding organizations of working in partnership is a coordinated, resilient and trusted workforce who understand the roles, responsibilities, capacity and capability of one another.
- When organizations work in partnership, people caught up in an emergency receive a joined-up, seamless, efficient and effective professional response from the agencies involved, appropriate to the situation.
- The key partnership for health organizations includes the Local Health Resilience Partnership. The membership comprises partner health organizations including the ambulance services and relevant local primary, secondary and tertiary care providers, Public Health England (or equivalent), local authority partners, as well as other non-health agencies in the local area.

Further Reading

British Red Cross (2015) Homepage. Available at: www.redcross.org.uk (accessed 26 November 2015).

Cabinet Office (2004) The Civil Contingencies Act 2004, Act and Guidance. Available at: https://www.gov.uk/guidance/preparation-and-planning-for-emergencies-responsibilities-of-responder-agencies-and-others (accessed 15 July 2016).

Cabinet Office (2013) The role of Local Resilience Forums: A reference document. Available at: www.gov.uk/government/uploads/system/uploads/attachment_data/file/62277/The_role_of_Local_Resilience_Forums-_A_reference_document_v2_July_2013.pdf (accessed 8 October 2015).

Department of Health (2012) Arrangements for Health Emergency Preparedness, Resilience and Response from April 2013. Available at: www.gov.uk/government/uploads/system/uploads/attachment_data/file/215083/dh_133597.pdf (accessed 8 October 2015).

Department of Health (2013) Health Emergency Preparedness, Resilience and Response from April 2013: Summary of the principal roles of health sector organisations. Available at: www.gov.uk/government/uploads/system/uploads/attachment_data/file/216884/EPRR-Summary-of-the-principal-roles-of-health-sector-organisations.pdf (accessed 29 October 2015).

HM Government, Department for Culture, Media and Sport and the Association of Chief Police Officers (2006) Humanitarian Assistance in Emergencies: Non-statutory guidance on establishing Humanitarian Assistance Centres. Available at: www.gov.uk/government/uploads/system/uploads/attachment_data/file/61221/hac_guidance.pdf (accessed 8 October 2015).

Joint Emergency Services Interoperability Programme (2015) Home page. Available at: www.jesip.org.uk (accessed 23 November 2015).

London Emergency Services Liaison Panel (2015) Home page. Available at: www.leslp.gov.uk (accessed 23 November 2015).

London Regional Resilience Forum (2006) Looking Back, Moving Forward: The Multi-Agency Debrief Lessons identified and progress since the terrorist events of 7 July 2005. Available at: http://news.bbc.co.uk/1/shared/bsp/hi/pdfs/23_09_06_lrrfreport.pdf (accessed 15 July 2016).

London Resilience Team (2016) London Prepared. Available at: www.london.gov.uk/about-us/organisations-we-work/london-prepared (accessed 23 February 2016).

National Commission on Terrorist Attacks upon the United States (2004) The 9/11 Commissioner Report. Available at: www.9-11commission.gov/report/ (accessed 8 October 2015).

NHS England (2015) Emergency Preparedness, Resilience and Response guidance. Available at: www.england.nhs.uk/ourwork/eprr (accessed 26 November 2015).

St John Ambulance (2015) Home page. Available at: www.sja.org.uk (accessed 26 November 2015).

Voluntary Sector Civil Protection Forum (2015) Home page. Available at: www.gov.uk/government/groups/voluntary-sector-civil-protection-forum (accessed 26 November 2015).

6 Interprofessional Working: Understanding Some Emotional Barriers and Unconscious Processes That Might Influence Practice in Group and Team Work

P. Sully

Honorary Visiting Researcher, Department of Psychology, University of Westminster, London, UK

> **Key Questions**
>
> - How can effective interagency team working be influenced by the nature of the emergencies to which the team attends?
> - What group processes might influence effective team functioning?
> - How might effective interagency teams develop further their best practice and thus learn from their shared working?

6.1 Introduction

The focus of this chapter is the exploration of some group processes that can influence effective interprofessional practice when working with colleagues from different agencies and disciplines in order to develop resilience strategies. It is also relevant when responding to civil emergencies and major incidents as interprofessional teams can be considered in four domains: relational, processual, organizational and contextual (Reeves *et al.*, 2010).

It is particularly relevant that for many years reports on responses to civil emergencies have underlined the vital importance of effective interdisciplinary working. Interprofessional cooperation is therefore essential throughout the planning, responding to and reviewing stages of managing civil emergencies (Clarke, 2001; Wapling *et al.*, 2009; Toner *et al.*, 2010).

6.2 Group Dynamics

The literature on group development and dynamics is extensive (e.g. Bion, 1961; Tuckman, 1965; Cartwright and Zander, 1968; Huffington *et al.*, 2004) and is underpinned by a number of theories. Psychodynamic and systems theories (Box 6.1) will be used here to outline some of the issues particularly relevant to this exploration of group and interprofessional team dynamics in the provision of services to people requiring them in civil emergencies.

Systems in organizations and teams are often given very rational explanations even though, to participants in the system, they might seem to militate against the primary task of the organization or individuals' roles in the delivery of the task. Where the work of the team involves service delivery in potentially life-threatening situations that can be very anxiety provoking, systems can develop to protect practitioners from intense anxiety although these systems/protocols/policies might be self-evidently counterproductive. Contexts of practice within individuals' inner worlds – professional and organization – influence how groups behave.

6.2.1 Overt and covert processes in teams

Relationships that are not directly involved in the interprofessional team nevertheless have an influence on the team dynamics as well as the processes in which practitioners work. These seemingly remote relationships, that can directly affect people in practice, can be forgotten to the detriment of competent and sensitive practice (see Hughes and Pengelly, 1997).

The manifestation of covert processes in teams is usually demonstrated through overt behaviour of which the meaning is not immediately evident; for example, sabotaging discussion by changing the subject, making a throwaway remark to detract from the focus of the work, whispering to a peer when

---

Box 6.1. Psychodynamic and systems theories: definitions

- *Psychodynamic theory* recognizes the dynamic nature of the human psyche, which includes both conscious and unconscious processes that influence human behaviour and interactions with others.
- *Systems theory* is based on the concept that the whole is greater than the sum of its parts. This approach recognizes that people's behaviour in groups and teams has conscious and unconscious elements. Human systems therefore include conscious and unconscious processes that can obstruct the group's or team's expressed purpose, or primary task.
- The *primary task* can be described as the clearly expressed purpose of the organization to which practitioners belong. Roberts (1994a) examines how this task is not necessarily that in which practitioners are directly engaged. She argues therefore that for teams to be effective it is essential that they clarify their primary tasks (see later).

These theories also recognize that management systems evolve to accommodate the conscious and unconscious needs of staff (Hoyle, 1994).

someone is putting a challenging point of view, or 'putting down' a colleague as a tease, but in a way that seems intended to humiliate. These types of behaviour can be described as *perversions* (Long, 2008; as cited by Ballatt and Campling, 2011) as they detract from the nature of the work and the common good. They flourish 'where people are used as a means to an end, as tools and commodities rather than respected citizens' (Ballatt and Campling, 2011).

This behaviour can detract from the nature of the work and may well be an unconscious response to feelings of inadequacy or apprehension. It also can lead to increased anxiety, less team trust and a sense that the project/ work of the team is not worth the effort. An example of this behaviour is where an individual or individuals go off the topic of discussion at a meeting or practice review, or raise irrelevant issues and discuss them at length, and consequently no decisions are made about the stated purpose for the meeting or practice review. This type of behaviour enables the team to avoid addressing difficult, painful or controversial issues.

The nature of resilience and emergency responses in a variety of arenas can mean dealing with a great deal of human distress. Where teams deal with suffering and challenge to their own abilities to function in the face of distress, they can share the defence of denial; that is, they collude with avoiding the trauma in the realities of their tasks. By the team sharing the denial of the suffering or pain they witness, members protect themselves emotionally. However, the risk is that the team loses its sensitivity by not recognizing the human reality of the situation and the suffering of those needing its services. They can thus deny survivors' anxiety or pain, as well as their own risk of pain as a result of the work.

Unconscious group responses to anxiety are identified in the work of Bion (1961) and developed further by others (e.g. Huffington *et al.*, 2004; Armstrong, 2005; Barratt and Campling, 2011). These responses are described as *basic assumption* mentalities that occur unconsciously in groups. These processes or mentalities (Bion, 1961) are the basic assumptions of:

- fight/flight (baF) – for example, where the group is in conflict to avoid (flee from) the task;
- dependency (baD) – for example, where the group relies on some authority who (or on some long-awaited protocol that) will magically rescue them/solve the problems inherent in the task (i.e. they are passive in the process); and
- pairing (baP) – for example, where they team relies on one or two people to make the decisions and manage the team by saying what each member needs to do.

In all these processes team members are not approaching the work from a position of their conscious adult experience and authority, but from unconscious processes and reasons.

Basic assumption behaviour can be anti-task; it can interfere with effective group functioning and thus delay, avoid and in some cases prevent the team from achieving its primary task. These responses are a result of the anxiety or threat perceived unconsciously in groups.

When groups or individuals feel anxious, even though they are not necessarily aware of it, they can also be drawn into 'performing a function on behalf of others' (Obholzer and Roberts, 1994). An example of this behaviour is a hospital first responder checking the ambulance emergency kit when this is the responsibility of the ambulance crew.

6.3 Interagency Planning

Emotion is a central part of the lives of groups and teams. Being aware of the above processes in organizational, team and individual communication can enable us to work more effectively and sensitively among disciplines; that is, to be an effective 'work group' (Bion, 1961; Armstrong, 2005). It is important too, to be aware of our own feelings in any given situation, as well as our own role and function in relation to the team's primary task.

Research cited by Hawkins and Shohet (2012) shows that when teams are first *forming* and *norming* (Tuckman, 1965) is the time when they need to focus on 'mission, goals and expectations of the performance of the team' (Hawkins and Shohet, 2012).

In managing team processes, whether as team leader or not, courtesy is crucial. Skills in assertively addressing group process when it is anti-task – such as keeping to agendas when they are being stalled, challenging discussions that are going over the same issues again and again even though resolutions have been agreed, or stopping meandering discussions that are unfocused and achieving little – are some ways in which team members can help each other to acknowledge the difficulties of certain aspects of their work and thus keep on task. Many of these manifestations of unconscious anti-task behaviour are summarized by Grasha (1995) as 'signs of hidden agendas'.

Focusing on *practice* and its development, and thus on *actions* rather than aims, is more likely to aid effective team working, 'expectations of performance' (Hawkins and Shohet, 2012) and what is expected of one another.

6.3.1 Conflicts of interest

Responders in civil emergencies work in their host organizations as well as members of interagency teams where the lead organization is not their own, so it is likely that at times they will have conflicts of interest where their host organizations may have differing priorities and/or policies. They may also find their loyalties are divided between the team's needs and the requirements of their employers (Roberts, 1994a,b). These situations can lead to difficulties in the running of the team, such as over timings and venues for meetings, resources available for service delivery and host managers' willingness to release staff at appropriate times.

Given that the need for effective interagency practice is central to the delivery of sensitive and appropriate responses to survivors and communities in civil emergencies, it is crucial that practitioners understand the importance of respectful relationships among each other. Working interprofessionally can be hindered by misunderstanding and assumptions about the relative positions and practitioners' purposes in service delivery (e.g. Obholzer and Roberts, 1994; Sully *et al.*, 2010). It is therefore essential that the purpose – that is, the *primary task* (Roberts, 1994a,b) – of the team be clearly identified (see Fig. 6.1).

Members of different disciplines are likely to have different perspectives on the issues central to service delivery. In the author's experience it is likely, however, that if the central issue/s are clearly stated (such as safety of staff and survivors/residents, rapid deployment of first responders), these perspectives are less likely to differ significantly from those of other team members.

Identifying and clarifying the primary task of the interagency team is, therefore, central to the process of effective planning and responding. Practice decisions in terms of priorities and how to achieve them are more likely to be agreed consensually. Where misunderstandings occur in practice at all levels, it could be the result of the primary task having not been clearly defined.

In the seminal work in on inter- and intra-agency working, *The Unconscious at Work: Individual and Organizational Stress in the Human Services* (Obholzer and Roberts, 1994), Roberts (1994a: 38) identifies the significance of the team in relation to the primary task. In order to consider effective individual and team working she suggests that it is important to ask the following:

- How does our way of working relate to this (primary) task?
- What are we behaving *as if* (italics original) we are here to do?
- How well are we doing?

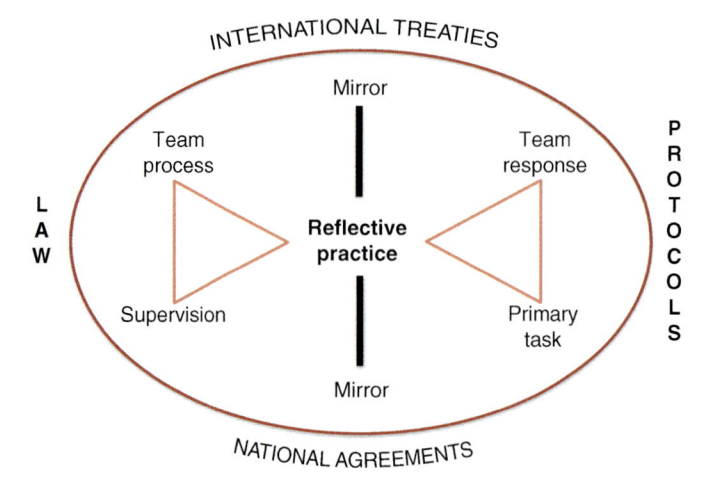

Fig. 6.1. The mirror as a catalyst for anticipatory reflection and reflection-in-and-on-action. © Sully, Wandrag and Riddell (2009). Published with kind permission of the authors.

She argues that where individuals are effective in teams we:

> need to be clear about the task we have to do, able to mobilize sufficient resources, internal and external, to achieve it; and have some understanding of how our own task relates both to the task of the system in which we are working and to the task of the institution as a whole.
>
> (Roberts, 1994a: 36)

Hawkins and Shohet (2012) discuss that interprofessional teams that have clearly articulated purposes, possess complementary skills and are able to negotiate priorities, depending on the circumstances and changing situations, are more likely to be high functioning (Obholzer and Roberts, 1994; Hughes and Pengelly, 1997).

Developing and maintaining effective team relationships can be challenging. Members can demonstrate, unwittingly, their or their organization's ambivalence about or indeed lack of commitment to involvement in the team, by behaviour such as not attending meetings, arriving late, choosing venues that are difficult to reach for some members, or not carrying out their agreed actions. Issues and processes such as these need addressing by the team and unreliable individuals followed up, if the team is to work competently and in a mutually respectful way.

6.3.2 Meeting structure and process

Cohen *et al.* (2011) identify the following characteristics of meetings that employees perceive positively:

- meetings start (regardless of whether all attendees have arrived) and end on time, or end before the due time;
- agendas are circulated prior to meetings to enable staff to prepare for them in advance;
- meetings are held in comfortable, appropriate environments with careful selection of attendees, so their presence is also significant; and
- informal agreements about behaviour in meetings are valued.

These findings can also be interpreted as treating colleagues with respect, setting boundaries or limits (e.g. clear time limits, agendas) and expecting colleagues' participation. Having clear parameters of engagement can make it easier to keep the primary task of the interagency team in focus. In these ways, anti-task behaviour can be limited but also respectfully addressed by others present.

Many of these observations seem self-evident. However, planning meetings sometimes need to be held at short notice, in unsuitable premises with little privacy or comfort, or as a response to an unforeseen event or management directive. Individual and/or team anxiety might thus be raised. In these circumstances covert relationships, perhaps between those 'in the know' or those who have closer relationships with influential people outside the team in their own or another organization, can have a direct effect on team dynamics and reactions to sudden changes. Team members who may appear to be adapting or responding positively, behind the scenes may sabotage the capacity of the group to develop creative responses to new demands.

Opportunities should be given to learn members' views and responses in meetings or in one-to-one reviews. This approach is not about intrusion into members' private feelings, but a sense of shared responsibility to deliver services that are responsive to community and individual need as opposed to being solely protocol driven.

Where organizations are target driven, the unexpected consequences to these changes can be a loss of sensitivity and humanity (see Francis, 2013, where targets were the priority over care delivery).

6.3.3 Reflection

The purpose of reflecting on practice is the further development of practice (Boud *et al.*, 1985). Individuals and teams can decide on actions to be taken by learning from experience (Wilson 2008; Sully *et al.*, 2010). Teams that review their work together honestly with mutual respect, despite the inevitable occasions where individuals feel irritated or disheartened, can build trust, develop and thus transform practice. Regarding differences in professional perspectives as strengths, rather than hindrances, is an important part of successful interprofessional and interagency working.

The processes that occur 'out there' are parallel to what goes on 'in here'. Groups and teams reflect in the meeting and practice review how they have delivered their services 'out there' (Sully *et al.*, 2008, 2010; Hawkins and Shohet, 2012). This mirroring process, if addressed as *part* of team process in facilitated review and debriefing, is regarded as a fertile source of learning, enabling teams to gain better insight into their behaviour, abilities and what supports or hinders them. In this way they can transform practice (Sully *et al.*, 2010). The mirror is a 'catalyst' (Sully *et al.*, 2010) for understanding the team's working processes, as well as its reflection upon what it may respond to in future. In the author's university experience, practitioners from a variety of disciplines found this model of reflection generally very helpful (Reeves and Sully, 2007). Experienced students who had participated in a rigid approach to debriefing, namely one that addressed solely the efficient and inefficient practicalities of managing an event/incident, commented that they found this approach more helpful. Reflection should never be used to blame.

6.4 Summary

That all people have unconscious processes that influence their behaviour is no less true of groups and teams. Interprofessional teams have much to offer in emergency preparedness, resilience, response and recovery. Their group processes and responses are influenced by context, the nature of the work and the anxiety it evokes, as well as the resources available at individual and organizational levels to deliver a service.

It is wise therefore to learn to identify signs of anxiety and consequential anti-task behaviour, as these are likely to interfere with efficient development and delivery of interagency emergency services.

Teams need time given to get to know one another and to learn together from practice, in an atmosphere of mutual respect. Clarification and evaluation of team process and service delivery is essential to team cohesion and team members' capacity to work together effectively in a variety of different circumstances.

The careful adjustment and modification of practice on scene and in later debriefing reflection supports effective interagency practice and its further development in preparing for, responding to and recovering from emergencies.

Skills in observation of team process, assertiveness and courteous challenge are essential in the clarification of roles, processes and tasks. Where teams have ill-defined primary tasks and little time to reflect together honestly, rather than to blame team members and/or others outside the group they can transform practice by learning from the mirroring process involved in teamwork in human services.

In these ways interprofessional and interagency teams are more likely to deliver appropriate and sensitive responses to those individuals and communities requiring them.

Key Answers

- Effective interagency team working can be influenced by the nature of the emergencies to which the team attends. Different emergencies invoke a different emotional response in both those caught up in the emergency and also the responders. Effective relationships between those affected by the emergency and those responding to it are reliant on the overt acknowledgement of these emotions. Equally, it is right to acknowledge the emotional response by those responders of different organizations.
- During the planning processes for emergencies individuals of different organizations will bring with them allegiances from their host organization. These overt and covert allegiances may have an impact on an interprofessional group's ability to remain on task and complete the preparedness project.
- An interprofessional group is more likely to succeed if the primary task is clearly articulated from the onset of the process and if relevant boundaries are set and maintained throughout. This framework therefore provides a safe structure to tackle difficult and emotional subjects and reduces the anxiety of those involved in the process.

Acknowledgements

With grateful acknowledgements to Kathryn Waddington and Malcolm Wandrag.

References

Armstrong, D. (2005) *Organisation in the Mind*. Karnac, London.
Ballatt, J. and Campling, P. (2011) *Intelligent Kindness: Reforming the Culture of Healthcare*. Royal College of Psychiatrists, London.
Bion, W.R. (1961) *Experiences in Groups: And Other Papers*. Basic Books, New York.
Boud, D., Keogh, R. and Walker, D. (1985) *Reflection: Turning Experience into Learning*. Kogan Page, London.
Cartwright, D. and Zander, A.F. (1968) *Group Dynamics Theory and Research*. Harper and Row, London.
Clarke, Lord Justice (2001) *Public Inquiry into the Identification of Victims following Major Transport Incidents*. The Stationery Office, London.
Cohen, M.A., Rogelberg, S.G., Allen, J.A. and Luong, A. (2011) Meeting design characteristics and attendee perceptions of staff/team meeting quality. *Group Dynamics: Theory, Research, and Practice* 15, 90–104.
Francis, R. (2013) *Report of the Mid Staffordshire NHS Foundation Trust Public Inquiry*. The Stationery Office, Norwich, UK.
Grasha, A.F. (1995) *Practical Applications of Psychology*, 4th edn. HarperCollins, New York.
Hawkins, P. and Shohet, R. (2012) *Supervision in the Helping Professions*, 4th edn. Open University Press, Maidenhead, UK.
Hoyle, L. (2004) From sycophant to saboteur – responses to organizational change. In: Huffington, C., Armstrong, D., Halton, W., Hoyle, L. and Pooley, J. (eds) *Working Below the Surface: The Emotional Life of Contemporary Organizations*. Karnac, London, pp. 87–106.
Huffington, C., Armstrong, D., Halton, W., Hoyle, L. and Pooley, J. (eds) (2004) *Working Below the Surface: The Emotional Life of Contemporary Organizations*. Karnac, London.
Hughes, L. and Pengelly, P. (1997) *Staff Supervision in a Turbulent Environment: Managing Process and Task in Front-line Services*. Jessica Kingsley, London.
Long, S. (2008) *The Perverse Organisation and its Deadly Sins*. Karnac, London.
Obholzer, A. and Roberts, V.Z. (eds) (1994) *The Unconscious at Work: Individual and Organizational Stress in the Human Services*. Tavistock, London.
Reeves, S. and Sully, P. (2007) Interprofessional education for practitioners working with survivors of violence. Exploring early and longer-term outcomes on practice. *Journal of Interprofessional Care* 21(7), 1–12.
Reeves, S., Lewin, S., Espin, S. and Zwarenstein, M. (2010) *Interprofessional Teamwork for Health and Social Care*. Wiley-Blackwell, Chichester, UK.
Roberts, V.Z. (1994a) The organization of work: contributions from open systems theory. In: Obholzer, A. and Roberts, V.Z. (eds) *The Unconscious at Work: Individual and Organizational Stress in the Human Services*. Tavistock, London, pp. 28–38.
Roberts, V.Z. (1994b) Conflict and collaboration. Managing intergroup relations. In: Obholzer, A. and Roberts, V.Z. (eds) *The Unconscious at Work: Individual and Organizational Stress in the Human Services*. Tavistock, London, pp. 187–196.
Sully, P., Wandrag, M. and Riddell, J. (2008) The use of reflective practice on masters programmes in interprofessional practice with survivors of intentional and unintentional violence. *Reflective Practice* 9, 135–144.
Sully, P., Wandrag, M. and Riddell, J. (2010) Supervision and facilitated reflective practice as central to disaster preparedness services to the older adult: a national and cross-national model. In: Toner, J.A., Mierswa, T.M. and Howe, J.L. (eds) *Geriatric Mental Health, Disaster and Emergency Preparedness*. Springer, New York, pp. 105–117.
Toner, J.A., Mierswa, T.M. and Howe, J.L. (eds) (2010) *Geriatric Mental Health, Disaster and Emergency Preparedness*. Springer, New York.

Tuckman, B.W. (1965) Developmental sequence in small groups. *Psychological Bulletin* 63, 384–399.

Wapling, A., Heggie, C., Murray, V., Bagaria, J. and Philpott, C. (2009) *Review of Five London Hospital Fires and Their Management, January 2008–February 2009*. NHS London, London.

Wilson, J.P. (2008) Reflecting-on-the-future: a chronological consideration of reflective practice. *Reflective Practice* 9, 177–184.

7 Command, Control and Communication

A. Rowe[1] and P. Thorpe[2]

[1]*Retired Metropolitan Police Operational Commander, London, UK; and Service Senior Emergency Planning Manager for the London Ambulance Service NHS Trust, London, UK*
[2]*Executive Director of the British Columbia Ambulance Service, Canada; previously Head of Olympic Planning for the London Ambulance Service NHS Trust, London, UK*

Key Questions

- What does 'command, control and communication' (C3) mean when considering an emergency or major incident?
- What is a 'concept of operations'?
- What command structures are recommended and commonly used?
- How are decisions and operations coordinated to resolve emergencies or major incidents?
- What is the role of control room staff when dealing with an emergency?

7.1 Introduction

Command, control and communication (C3) for leadership and management are as important in the field of health as in any other organization that has to manage a response, whether generated externally or internally.

This chapter briefly outlines the system of C3 that generally operates within the UK including its definitions: the 'concept of operations', strategic, tactical and operational leadership and management, decision making, and communications and record keeping within the control room and elsewhere.

Throughout this chapter the London 2012 Olympic and Paralympic Games (hereafter 'the Games') are used to illustrate the concepts discussed within planned events, against the background that there was always the risk of additional rising-tide events, such as an outbreak of infectious disease or a spontaneous event such as a terrorist attack or stand collapse (see Box 7.1).

©CAB International 2016. *Health Emergency Preparedness and Response*
(eds C. Sellwood and A. Wapling)

> **Box 7.1.** Definitions of incident types
>
> - *Rising tide*: developing from a steady state to becoming an emergency or major incident over a more prolonged period of time (e.g. influenza pandemic).
> - *Planned*: where there has been an opportunity and time to develop strategies, tactics and contingencies before an anticipated or known event takes place.
> - *Spontaneous*: (often referred to as 'big bang' or 'rapid onset emergency') where there has been no prior warning to develop strategies, tactics or plans. In spontaneous events a generically planned initial response may be directed towards interventions to prevent escalation. Such emergencies may include transportation incidents, adverse weather or natural disasters, or acts of terrorism.

The interagency C3 structure employed during the Games was developed to be flexible in its response to any of these types of event while ensuring that the planned event (the Games) would continue.

The overriding fact to successfully concluding an emergency, having already produced a plan that has been trained, exercised and revised as necessary, is to ensure that the organizational management structure provides clear leadership, accountable and defensible decision making, as well as up-to-date and far-reaching communication (both internal and external). This structured approach to leadership and management under pressure is commonly known as 'command and control'.

7.2 Definitions

There are numerous definitions of C3. The summaries shown below are for reference purposes and to provide context for this chapter.

- *Command*: the ability to give an order/instruction requiring action; requires/demands attention and obedience (authoritative/peremptory). (Note: one can give orders without being 'in control'; see below).
- *Control*: the power to influence or direct people's behaviour and/or actions or the course of events, and the methods used to exercise both command and control.
- *Communication*: the passing and receiving of information, in this instance in a purely strategic and operational environment (not to be confused with 'communications' as discussed in Chapter 8, this volume).

In a multi-agency environment there may be a number of agencies with compatible interests but differing governance structures and objectives. In these circumstances there may be a stronger focus on the development of coordination and communication between agencies than that of command and control. During the Games partner agencies were co-located at the National Olympic Coordination Centre (see Case Study 7.1), allowing for an integrated multi-agency approach to the delivery of a safe and secure Games.

> **Case Study 7.1.** The National Olympic Coordination Centre
>
> During the Games, the National Olympic Coordination Centre (NOCC), while an integral part of the C3 structure, had no direct command of any national or local resources. It was staffed by partner agencies: Games organizers, fire, police, ambulance, the Greater London Authority, local government and others, under the direction of the National Olympic Security Coordinator. The Health Desk in the NOCC was staffed by senior leaders from the ambulance service with a direct link to the Department of Health (DH) Situation Room. During the Games the NOCC Health Desk provided updates to DH who could then give direction through local coordination centres and control rooms to health care providers.

It is worth remembering at this early stage that these three definitions cannot stand alone and have to be considered against a wider background that includes: *intelligence–coordination–collaboration–communications* (in the sense of Chapter 8, this volume).

This book has so far dealt with matters relating to risk assessment, planning in partnership and interagency working in the operational context. The culmination of planning, training, exercise and review all comes together under the operational umbrella when an emergency or major incident is declared.

While multi-agency cooperation when dealing with emergencies has taken place over a number of years, the Civil Contingencies Act 2004 codifies the UK requirements for agencies to work more closely together in all elements of planning, training, exercise, review, response and recovery. The immediate strength of the Act was to require civil authorities and other agencies (e.g. public utilities) to work towards a common goal under a universal command and management structure that was understood by all, without the risk of confusing the roles of organizations or personnel involved by overlap and/or working in silos, without reference to each other, etc. It also encourages organizations (referred to as Category 1 and Category 2 Responders in the Act) to use the same management processes when dealing with internal emergencies.

This management concept has recently been further endorsed by the emergency services in the UK when the Government (the Home Secretary) required the three emergency services to conduct joint training that endorsed the processes outlined within the Civil Contingencies Act 2004 and associated guidance (Joint Emergency Services Interoperability Principles (JESIP)).

The approach to exercising of the C3 capacity for the Games involved all partner agencies and provided assurance to stakeholders of games readiness (see Case Study 7.2).

It is also vitally important to note that in addition to formal exercises, the command and control structures used to deal with emergencies in the UK are the same as those used on a regular basis by multi-agency teams dealing with various pre-planned events. These range in scale from the Games and royal weddings, through sports events and pop concerts, to local fetes and fairs.

> **Case Study 7.2.** Interagency emergency management during the Games
>
> Interagency emergency management during the Games was complex, with consideration of the needs of all agencies involved having to be taken in to account. This led to significant investment of time and resources into the planning of the Strategic Testing and Exercising Programme (STEP) led by the UK Government Home Office through the Olympic and Paralympic Security Directorate as well as local and national testing.
>
> A number of national events were held including Operation Amber, which tested all elements of the UK Ambulance Service response to the Games including C3, pre-planned aid, the Olympic Deployment Centre, clinical training, specialist multi-agency response and national communications networks. This was undertaken as part of the assurance process for the Association of Ambulance Chief Executives, DH, Home Office and other agencies.

7.3 Concept of Operations

Prior to such event plans being drafted, the overarching structures and intentions that support an event need to be identified by the principal organizers in partnership with interagency partners. This will lead to the jointly agreed production of a document known as a 'concept of operations' (CONOPS). Such a document will broadly set out agreed goals and overarching principles. For example, the London Olympic CONOPS set out (among other things) the following six aims and objectives:

- *roles and responsibilities* – decision makers are clear on their roles and responsibilities and their delegations and limits of authority;
- *information and communication flows* – there are clear information flows into and across the C3 to ensure all decision making is informed by a common understanding of the overall situation;
- *timely and informed decision making* – there are adequate and resilient arrangements in place for consultation on major issues;
- *supporting infrastructure* – there are adequate facilities and communications links to support decision making and disseminate those decisions appropriately (potentially the establishing of a joint multi-agency control facility, and other cross-organizational links, e.g. teleconferencing);
- *communications* – public messages issued by all those responsible for the delivery of the Games are consistent and coordinated; and
- *response to emergencies* – the C3 is designed to support the response to both Olympic and non-Olympic emergencies.

In support of these principles, each organization that had a role relating to the Games was expected to produce its own CONOPS that supported the national concepts. These organizations included a remarkably wide range of partners, including:

- Games organizers at each venue and residence (hotels);
- police;
- fire;

- ambulance;
- local designated hospitals;
- local authorities (to include the special requirements of their local community);
- transportation companies;
- public utilities and volunteering organizations;
- organizers of cultural events; and
- many others against a background that required 'business as usual' to be maintained.

While the various CONOPS documents were owned by the various individual agencies involved, they were peer reviewed internally and externally against the six criteria outlined above. Each was required to produce a risk assessment and a plan that identified its own risks. Subsequently individual and later interagency plans could be produced that ensured all perceived threats and challenges were considered and mitigated to the best of their corporate ability. The operational plans also ensured that an adequate response could be made by each organization, working with partners, to reduce, control or mitigate the threats posed or resolve any emergency that may have been declared. Thus once an emergency had been declared by an organization (or internally by a department within an organization) other partners would move to provide the necessary support by way of initial response leading towards recovery.

7.4 Command Structures

Each organization (or department when dealing with an internal incident) will follow the same basic template for planning, command and control that will be required to respond effectively and thus prevent, mitigate or resolve both external or internal emergencies and threats, with the intention of concluding the emergency and restoring normality as quickly as possible in a planned and coordinated manner, always against a background of maintaining (as far as possible) 'business as usual'.

The template is based very simply on the concept of three tiers of management: *strategic*, *tactical* and *operational*. These have been adopted by each of the emergency services as well as (within London) Category 2 responders as identified by Civil Contingencies Act 2004, including local government, public utilities and others. These are each discussed in more detail below. The Games' original Bid Document outlined responsibilities, including those of health care providers, in the event of an incident during the Games (see Case Study 7.3).

7.4.1 Strategic

The purpose of the strategic level (often referred to as 'gold' by individual responder agencies) is to consider the emergency in its widest context:

- determine longer-term and wider impacts and risks with strategic implications;
- define and communicate the overarching strategy and objectives for the response;

> **Case Study 7.3.** Command structures at the Games
>
> The London 2012 Games management process required each of the medical services to retain clear responsibility for the deployment, command and control of its resources using the procedures outlined in the London Emergency Services Liaison Panel (LESLP) Manual. This then included the use of the 'gold, silver and bronze' command structure commonly referred to as 'GSB'. JESIP 2014 now recommends the use of 'strategic–tactical–operational' tiers of management.

- establish the framework, policy and parameters for lower-level tiers; and
- monitor the context, risks, impacts and progress towards defined objectives.

Therefore, from the outset of the incident response, each strategic manager will determine his/her organizational intention and record a strategic statement for his/her own organization. This statement will need to be monitored, subject to ongoing review and updated as required by the interagency Strategic Coordinating Group (see below).

7.4.2 Tactical

The purpose of the tactical level (often referred to as 'silver' by individual responder agencies) is to develop interagency procedures that support the strategy by ensuring that the actions taken by the operational tier are coordinated, coherent and integrated in order to achieve maximum effectiveness and efficiency.

The tactical manager may attend the scene of an incident, take charge and be responsible for formulating the tactics to be adopted by his/her organization to achieve the strategic objectives set by the strategic manager. The tactician should not become personally involved with activities close to the incident, but remain detached.

There will be occasions when that person may be based elsewhere and a scene manager (see below) will fulfil the function normally undertaken by the tactician. This may occur under various circumstances including multiple simultaneous scenes, threats or incident sites (e.g. several explosions, or marauding armed terrorists) where a coordinated tactical response across sites is required, or when an incident occurs where a command structure is already in place for a pre-planned event or operation (e.g. a major incident during a sporting event or concert).

7.4.3 Operational

Operational is the level at which the management of immediate 'hands-on' work is undertaken at the site(s) of the emergency or other affected areas. Individual responder agencies may refer to the operational level as 'bronze'.

Operational managers will control and deploy the resources within a geographical area or to a specific role to implement the tactics formulated by the tactician. Clearly there will be as many operational managers as required to resolve the emergency, each of whom will report to their own tactician while liaising with operational managers from other organizations or departments as required. At an internal incident, such as a hospital power failure, these will be the relevant managers from any department that has a role in controlling or mitigating the threat posed; for example, those dealing with IT, bed management, wards, catering, facilities, the emergency department, etc. It is vitally important that tactical and operational managers are easily identifiable to each other (e.g. by the use of high-visibility jackets or tabards) and other personnel.

The strength of this agreed structure is that it ensures organizations and departments understand each other's roles, ability and authority. However, it remains vitally important to understand that any organization may request temporary assistance, personnel or equipment from another. Essential to the success and confidence to do this is to ensure consistent competencies across these functions (see Case Study 7.4). In such circumstances, while the supporting service will relinquish the immediate control of those resources to another for the duration of the task, it will nevertheless keep overall command of its personnel and equipment at all times. The rationale being that personnel from one service or department who help another in this way should only be given tasks for which they are trained and not simply supplement the other service in a potentially dangerous situation, and that knowledge is clearly best known and understood by the providers' management and leadership.

Case Study 7.4. Command and control at the Games

Performance criteria were established that determined a number of competency standards relating to the degree of knowledge and skills required to operate at the Games; these levels are contained in the National Occupational Standards. Each partner organization in the NOCC undertook to train its command structure to these levels:

- responding to emergencies at the strategic (gold) level;
- responding to emergencies at the tactical (silver) level; and
- responding to emergencies at the operational (bronze) level.

The ability to work at these levels was tested during the STEP in command-post and live exercises.

7.4.4 The Strategic Coordinating Group

A primary task following confirmation that a multi-agency (inter-departmental) emergency has or is likely to occur is the establishment of the multi-agency Senior Management Group, which is often known as the Strategic Coordinating Group (SCG). The SCG involves strategic managers with the appropriate mix of seniority and authority from each responding organization or department.

A multi-agency SCG in the UK may be chaired by a senior police officer (the police have the role of coordinating the response to a wide variety of incidents including terrorism, transportation incidents and natural disasters), but this does not place the police service 'in charge' or 'in command' of the resolution of the emergency or other resources unless by agreement.

Having received all the relevant information from each organization involved and thus its strategic approach to resolving the emergency, the SCG will produced an agreed overarching Strategy that will drive the resolution forward, taking care to merge and blend with the strategic approach adopted by the other organizations or departments.

Clearly an internal emergency, not requiring the involvement of other agencies, is advised to follow the template outlined in this section but the strategic lead will be provided by the most appropriate person within the organization with the necessary degree of authority, skills and training to achieve the desired result.

It is vitally important that an SCG is established at the earliest opportunity, by whatever is the most convenient means including face to face, telephone or video conferencing. Here each organization's strategy can be shared and a single, simple, overarching Strategic Statement can be agreed. This will determine the immediate priorities and may, even at this early stage, consider the transition from the response phase into recovery. SCGs during the Games were established in time for the STEP programme and were activated prior to the Torch Relay (see Case Study 7.5).

Case Study 7.5. Strategic Coordinating Groups during the Games

During the Games a number of standing SCGs were established in key areas, such as London and Dorset. These groups met under the chair of the police force with responsibility for the security in those areas. Groups would be convened daily to review areas such as the order of play for the day, VIP or protected person visits, latest intelligence and any agency-specific issues. These groups enabled the flow of information into each agency in a steady state.

In the event of any emergency or incident, the infrastructure and routine for the SCG were already in place. Core membership of agencies such as police, fire and ambulance would be supplemented as appropriate in such a case. A key factor in planning for these groups was the ability to function for an extended period of time, especially if an incident occurred in the early stages of the Games, such as the Torch Relay.

SCG meetings must then take place at agreed regular intervals, usually against a fixed agenda. The first item is typically 'Urgent decisions required', followed by 'Updates' (including 'Progress reports' and 'Fresh challenges') from all participants and the agenda concludes with an assessment of progress against the Strategy in order to drive tactics for the next phase or period of time.

During the course of the emergency, and thus at the SCG meetings, it is necessary to continually review the corporate risk assessment, by embracing the known or perceived risks from the organizations involved. This, together with horizon scanning (looking forward) and a review of progress, may require that the Strategic Statement be adjusted or amended during the resolution of the emergency as progress is made or new threats and challenges are identified. All this highlights the vital need for all such meetings and decisions to be minuted comprehensively and circulated quickly. This formal activity should ensure that the decisions taken are in support of the Strategy and are corporate and defensible, both individually and organizationally at any post-emergency enquiry.

Each member of the SCG armed with the agreed Strategy and decisions taken MUST then inform his/her tactical manager as quickly as possible in order that the necessary tactics can be designed, personnel and equipment requested and provided so that implementation can be initiated to drive the Strategy, and thus the resolution of the incident, forward.

Strategic statements need not be lengthy but must provide the agreed intention of the organizations to drive both response and recovery. For example, at a transportation incident the SCG's strategic statement might be as shown in Box 7.2.

This simple statement involves all the organizations that may be expected to deal with response to an emergency: fire, police, ambulance, local government, transportation authorities (road, rail, air) and hospitals. Each will develop tactics that will support the statement. It is reasonable to expect that each organization will be generally aware of the ability of other agencies and as the result of familiarization through joint planning, training, exercise and review. Similar simple strategic statements can be drafted that will effectively drive internal incidents ranging from loss of electricity or IT to the need to evacuate.

Subsequent to any strategic meeting it is vital that the tactical managers are informed of the decisions taken and that they hold regular meetings with interagency (or inter-departmental) tactical colleagues. This group, now commonly known at the Tactical Coordination Group (TCG), will discuss priorities and actions to ensure greater consistency and management of what can be achieved, with the result that their

---

Box 7.2. Strategic statement for a transportation incident

Make safe the area to ensure:

- a safe working environment for rescuers; from which
- the injured can be removed as quickly and safely as possible to appropriate designated hospitals;
- appropriate facilities can be made available for other survivors;
- the deceased can be dealt with sensitively and identified;
- enquiries from friends and families can be dealt with effectively;
- investigation of the circumstances can take place; and
- normality can be restored as quickly as possible.

tactics can complement each other in the best interests of supporting the Strategy and resolving the emergency.

Tactical managers can then brief their operational managers as to the way forward, the latter being responsible for deploying the necessary resources. It is essential that regular reports of progress (situation reports or SITREPS) are forwarded (to and from) through the chain of command. This will ensure the strategic leads are aware of progress and changes so that provision can be made for either additional appropriate resources or a planned and seamless move from response to recovery.

7.5 Communication

None of the above can be achieved without adequate communication between individual organizations or departments and, just as importantly, with other organizations involved, as previously stressed. Indeed, communication is the very essence of 'control'. Throughout planning for the Games significant effort was put into the development of common language (see Case Study 7.6) for use by organizations to support decision making.

Control begins through the strategic manager, who should have a support team able to deal with high level (e.g. central government) and media enquiries, obtain up-to-date information and intelligence from a wide range of sources, and coordinate progress. Constant horizon scanning must take place to ensure that the right equipment and personnel are available at the right place at the right time and that proper arrangements are made for the welfare of personnel deployed. Information from the scene and elsewhere would ensure that the correct decisions are made at the right time. Here again it is a requirement that information, intelligence and routine messages are recorded as each forms part of the operational log from which reports can be prepared and statements taken. All documentation, recorded in any control facility and from whatever source (electronic and physical), will need to be saved and produced at any subsequent internal or interagency debrief, formal enquiry, civil or criminal proceedings, or coroner's court.

Of equal importance is the need for all managers to record, as far as they are able, the rationale for decisions taken, including, if possible, the options considered. Details of any advice sought and received should also be

---

Case Study 7.6. Interagency communications during the Games

Interagency communications should be clear and concise. Discussions were held pre-Games on what this meant including the type, frequency and route of sharing of information. This included considerations on the confidentiality of information shared between organizations. Other factors considered included the interoperability of systems and reach back to parent organizations. Finally, an understanding of the different usage of common terms was reached. For example, a 'patient' to health organizations means someone injured, treated and/or transported to a place of treatment. To other agencies this term was being used to describe a member of the public involved in an incident.

recorded. It is however recommended that this strategic group is separated from but situated adjacent to the control centre. Such a separation tends to ensure that information is passed to the most appropriate levels and not necessarily referenced to the strategic group just because it is co-located.

7.5.1 Control rooms

There is almost always a need for each organization to operate from its own control environment, as well as probably being represented at the corporate control room. Many control rooms are pre-established, with specially trained and exercised staff that are dedicated or can be withdrawn from their normal duties to staff the room. Each person must clearly understand his/her role and the function he/she provides for emergency response. As far as possible (and this is recommended), liaison personnel from participating organizations or departments should send representatives to each other's control facilities to aid interagency communication and liaison.

In the development of plans for control rooms, consideration must be given to the interoperability of IT systems, interaction with other control rooms, the need to set up at short notice, especially in spontaneous incidents, and potentially the need to operate for extended periods of time.

Clearly, control room staff cannot be parachuted into a control room without adequate training and exercising. This is especially true if personnel are to be drawn in from their day role. One of the primary considerations in ensuring that an emergency is dealt with or mitigated as quickly, efficiently and effectively as possible will be the performance of control room personnel. Not only must they have a comprehensive knowledge of the organization's expectation of them, but a full understanding of the incident and its progress towards resolution, including pre-deployment briefings and ongoing updates.

7.6 Summary

This chapter has outlined the overarching processes that have been used in the UK for a number of years. The process is dynamic and changes are constantly made as the result of training, exercising, events and emergency response. The processes identified above, with good leadership and sound decision making, have stood the test of time within challenging environments and against a wide variety of large-scale events, real threats and emergencies.

This chapter has identified the advantages of relationships between and integration of C3 in terms of single organizations (diverse departments) and multiple agencies. It has also identified the fact that such processes, when coherently structured, planned, drafted, exercised, tested and reviewed, can be used as a response to a variety of challenging incidents whether anticipated or spontaneous, from small-scale to major events, best exemplified by the Olympic Games, natural disasters, transportation and terrorist incidents, and a variety of health emergencies.

> **Key Answers**
>
> - Command, control and communication (C3) – together with other elements including intelligence, coordination, collaboration and (media) communication – is the acknowledged methodology adopted in the UK to deal with the response and recovery phases of an emergency. It is flexible and can be used in a multi-agency environment or within a single organization that is dealing with a significant incident involving more than one department.
> - The concept of operations can be drafted ahead of a pre-planned or anticipated event or emergency. This should be tested and agreed by the various organizations (or departments) involved. This will lead to roles and responsibilities being understood; information and communication flows established; and informed and defensible decisions being made. This is all against a background that ensures that supporting infrastructure is adequate and communications (internal and external) are established. The overarching key to success is the ability to respond to a variety of specific challenges against a 'business as usual' background.
> - The Civil Contingencies Act 2004 guidance document on Emergency Response and Recovery has identified best practice in recommending the use of strategic, tactical and operational tiers of management, supported by a strong infrastructure within each organization as well as corporately (interagency) by way of the Strategic Coordinating Group (SCG), now augmented by the recent Joint Emergency Services Interoperability Principles (JESIP).
> - Coordination is maintained through the SCG, formal minutes, and decisions being swiftly communicated to the tactical tier for implementation. This is supported by constant risk review, horizon scanning and reviewing of progress.
> - Control room staff are one of the principal keys to resolving an emergency. Whether dedicated or drawn into the control facility, they must be trained and exercised for specific roles, kept fully updated, and aware of the importance of their role in maintaining the log of events and thus their role in debriefs and any further enquiry.

Further Reading

Cabinet Office (2006) Emergency preparedness. Available at: www.gov.uk/government/publications/emergency-preparedness (accessed 8 February 2016).

Cabinet Office (2013) Emergency response and recovery. Available at: www.gov.uk/emergency-response-and-recovery (accessed 8 February 2016).

Joint Emergencies Interoperability Programme (2015) Home page. Available at: www.jesip.org.uk (accessed 8 February 2016).

Mayor's Office for Policing and Crime (2015) London Emergency Services Liaison Panel. Available at: www.leslp.gov.uk (accessed 8 February 2016).

National Ambulance Resilience Unit (2016) Home page. Available at: www.naru.org.uk (accessed 8 February 2016).

National Occupational Standards (2016) Home page. Available at: http://nos.ukces.org.uk/Pages/index.aspx (accessed 22 February 2016).

NHS England (2016) Emergency Preparedness, Resilience and Response (EPRR). Available at: www.england.nhs.uk/ourwork/eprr/ (accessed 8 February 2016).

Public Health England (2016) Home page. Available at: www.gov.uk/government/organisations/public-health-england (accessed 8 February 2016).

8 Communications During a Health Emergency

J. Cole

Senior Research Fellow, Resilience and Emergency Management, Royal United Services Institute, London, UK

Key Questions

- Why do you need to communicate?
- Who do you need to communicate with?
- When do you need to communicate?
- What do you need to communicate?
- How can you communicate effectively?

8.1 Introduction

A robust and resilient communications strategy is especially important during health emergencies. The spread of a disease can be slowed and eventually contained by promoting and enabling effective infection control, facilitating surveillance and contact tracing of known cases, and ensuring that timely and scientifically accurate information is disseminated to appropriate groups within effective time frames. The World Health Organization (WHO) *Outbreak Communication Planning Guide* (www.who.int/ihr/elibrary/WHOOutbreakCommsPlanngGuide.pdf) states:

> pro-active communication encourages the public to adopt protective behaviours, facilitates heightened disease surveillance, reduces confusion and allows for a better allocation of resources – all of which are necessary for an effective response.

This is illustrated in Fig. 8.1.

During the recent Ebola crisis in West Africa, effective communication strategies have been credited with limiting the spread of the disease from early cases in Senegal and Nigeria. Figure 8.2 illustrates some public messaging in Sierra Leone in 2015.

Good communication strategies can also help to mitigate the health impacts of contamination events, for instance by: (i) informing people to stay

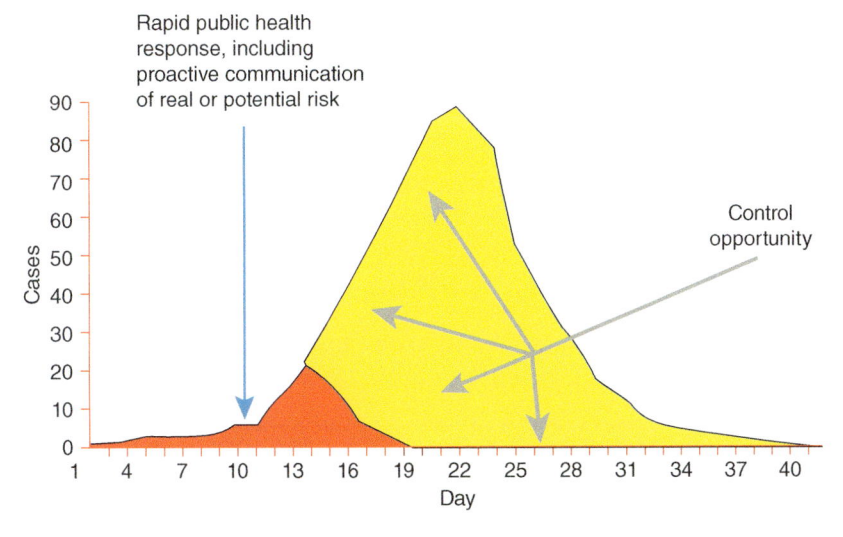

Fig. 8.1. Proactive communication in infection control. (From www.who.int/ihr/elibrary/WHOOutbreakCommsPlanngGuide.pdf.)

Fig. 8.2. Roadside public health messaging in Sierra Leone, 2015. (© Tom Mooney.)

inside if a chemical fire nearby has polluted the air; (ii) giving early warnings to a local population of extreme heat, cold or other weather events that may affect their health, so that they have time to plan for them; and (iii) ensuring that people are aware of support that is available, including long-term health monitoring programmes, following an event.

8.2 Health Communication: the Basics

Michael Hallowes, the then-head of the UK National Policing Improvement Agency, once stated that communication is 'not just the C of ICT (Information and Communications Technology)'. Equally important as the technology aspect is knowing with whom to communicate and how to frame the message. In the case of a health emergency, this will include communicating sometimes complex scientific and medical information to emergency responders, the media and the public in a way that can be easily understood. Effective communications will encompass:

- incoming communications and information from outside the organization, including from organizations whose involvement may be specific to the event;
- internal communications between an organization's own workforce; and
- external communications to collaborating agencies, the media and the public.

Messages need to contain not only facts and figures on, for example, how a disease spreads and how many people have currently contracted it, but also risk communication, including information on what the risk is, how to avoid it or mitigate its effects, the timescale of the risk and when it may increase or decrease. Who is responsible for managing the risks should also be clear, so that everyone knows what their responsibilities are and who else they may be responsible for.

Communications should manage expectations, be honest and practical, and be prepared to explain why some information may be unknown, uncertain or liable to change. Being unsure of the current situation and how it might develop is not an excuse for giving no communication at all – in an information gap, rumours and misinformation are more likely to spread if there is nothing and no one to counter them. The WHO *Outbreak Communication Planning Guide* (www.who.int/ihr/elibrary/WHOOutbreakCommsPlanngGuide.pdf) has set out five important principles of communication during health emergencies:

- trust;
- announcing early;
- transparency;
- listening (to the public's concerns); and
- planning.

Two particularly good online tools for helping to plan communications strategies during health emergencies have been developed by the International

Federation of Red Cross and Red Crescent Societies (IFRC) (www.eird.org/esp/ifrc-toolkit/guia/emergency-communications.pdf) and the European Commission (http://ec.europa.eu/health/preparedness_response/docs/gpp_technical_guidance_document_1_december_2009.pdf).

8.3 Timeline of Communication

Guidance issued by the UK's Civil Contingencies Secretariat (CCS) clearly sets out the stages of an emergency and gives guidance on each stage in its publication *Emergency Preparedness* (Fig. 8.3).

Although the CCS Guidance is intended for generic emergencies, it can easily be applied to specific health emergencies (see www.gov.uk/government/uploads/system/uploads/attachment_data/file/61030/Chapter-7-Communicating-with-the-Public_18042012.pdf):

> Advance preparation is essential and developing an outline communication strategy to deal with incidents is vital. When an incident/emergency occurs, this rolling strategy can be quickly developed to ensure a comprehensive and coordinated strategic communications approach is taken. This strategy however should continue to evolve over the duration of the incident to ensure that the most effective and appropriate action continues to be taken. Where possible, there should be a range of communications specialists involved in developing the strategy so that all communication disciplines, from free to paid media, e-media to direct mail, are considered.

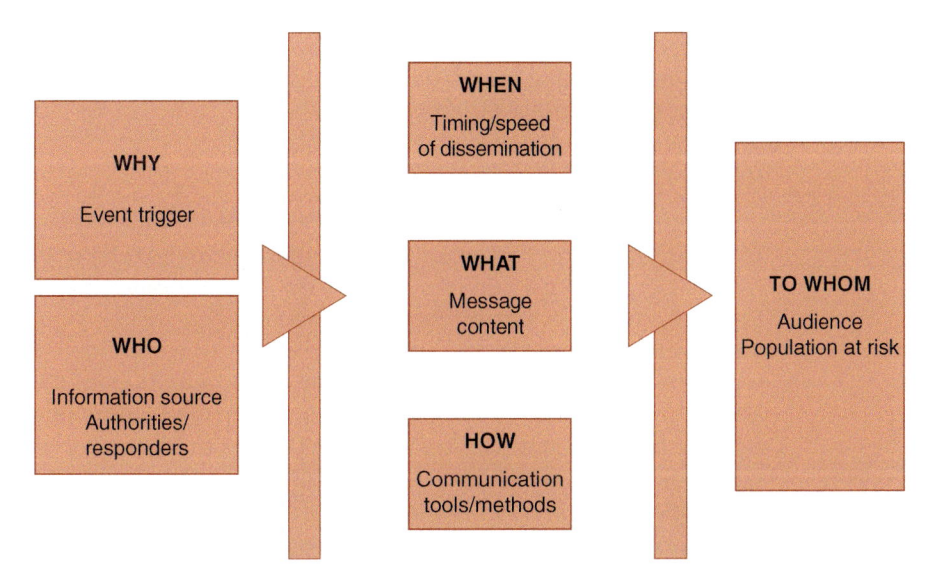

Fig. 8.3. Communications process. (From Chapter 7 in Revision to Emergency Preparedness, Civil Contingencies Enhancement Programme, Cabinet Office, 2012. Available at: www.gov.uk/government/uploads/system/uploads/attachment_data/file/61030/Chapter-7-Communicating-with-the-Public_18042012.pdf.)

8.3.1 WHY: event trigger

In the event of a health emergency, the event trigger may be early identification of cases of a disease (such as measles or influenza), suggesting a localized or potential outbreak, or a number of individuals with unusual symptoms self-presenting at a hospital or health centre, which may be indicative of food poisoning or a chemical attack. The trigger may be related to an international event, such as the outbreak of Ebola in West Africa in 2014, or the nuclear disaster at the Fukushimi Dai-ichi nuclear power plant in Japan in 2011, that is not of immediate danger to the UK but may require liaison with the media to ensure news coverage is balanced and does not cause undue concern. In the case of the 2009/10 swine flu pandemic, early identification of cases in Mexico enabled the UK to prepare a communications strategy to health care workers and the public well in advance of cases reaching the UK (see Chapter 14, this volume).

8.3.2 WHO: information source – authorities, responders

In today's modern communications environment, where mobile phones and Internet-connected devices are everywhere, information can come immediately from multiple sources across multiple platforms.

During a health emergency, it will be important to ensure that information coming from international, national and local sources is consistent and gives the same advice; or, if different advice is given by different organizations for good reason, this is explained so as not to cause confusion. Workers in non-governmental organizations based in Liberia during the 2014/15 Ebola crisis reported that they felt most assured when the information given by the Liberian Ministry of Health, the WHO, international news media such as the BBC and local media sources was consistent.

Information sources will include:

- international health organizations such as the WHO, the US Centers for Disease Control and Prevention (CDC) and Medicins Sans Frontiers (MSF);
- national health organizations such as the Department of Health, NHS England and Public Health England in the UK;
- international, national and local media;
- social media, which can be used to monitor the situation on the ground and help to understand the public's reaction and concerns (the Association of International Air Transport (IATA) has produced a particularly good guide to social media in crisis communications; see www.iata.org/publications/documents/social-media-crisis-guidelines.pdf); and
- personal communications from friends and colleagues directly involved in the event.

Organizations that are likely to be issuing official communications and statements need to liaise with one another to ensure the messages they are giving are consistent.

8.3.3 WHEN: timing/speed of dissemination

While modern technology means that messages can be sent across the world instantly, it still takes time to formulate and send the right message; that is, one that contains accurate and appropriate information. Preparing such messages may take time and, in the interim, holding messages can provide assurance that the situation is coming under control and that further information will be provided when available. Social media makes it inevitable that information will be pushed out extremely quickly from members of the public on the ground – for example, by people working at a hospital where early cases of a serious disease outbreak are being treated – so it is important that official messages are issued just as quickly so it does not look as if anything is being 'covered up'. The WHO warns that:

> The longer officials withhold information, the more frightening the information will seem when it is eventually revealed, especially if it is revealed by an outside source. Late announcement will erode trust in the ability of public health authorities to manage the outbreak.

It is also important to consider the temporality of information – how long is this valid for and when can those receiving it expect to be given an update? More information is given in Table 8.1.

8.3.4 WHAT: message content

The message content does not need to contain perfect information: the best estimate of the current situation will help to prevent rumours and misinformation spreading and stop accusations that the authorities are 'doing nothing'. Message content should contain the facts as far as they are known and explain these clearly to a non-expert audience. For instance, in a health emergency it may be important to explain the difference between figures given for confirmed and suspected cases; and to use non-jargonistic terms. The information 'the background radiation in Tokyo is 66 nSv/h' does not mean much to most people. A better way to phrase it is to add 'this is higher than usual for Tokyo at this time of year, but less than it would be during a summer day in the South of France'. Framing a message in such a way helps people to understand how they may be affected.

The advice from the UK CCS is that:

> the more information the public has access to, and the better educated they therefore become before an event, the more open they are likely to be to the warnings and advice they are given at the time of an emergency.

It is also important to be clear on why some information is being withheld from the public, if a decision to do so is made, and to be prepared to explain the reasons for this decision if the information is 'leaked' at a later stage.

Table 8.1. Timing communications: before, during and after a health emergency.

Before	• Be aware of what risks may result in a health emergency, including infectious disease outbreaks; extreme heat or cold spells; contamination from chemical fires or radiological incidents; food poisoning and others
	• Know what the early warnings of an event are likely to be, where notification is likely to come from, and who you are responsible for communicating messages and information to
	• Know who is responsible for what aspects of the emergency and how to communicate with them when an event occurs
	• Write detailed but flexible plans, which can form the framework of a response
	• Ensure appropriate training is provided and exercise the plan regularly
	• Know who needs to be informed and how they can best be reached, including vulnerable populations such as the elderly, disabled and those who may face language or literacy challenges
	• Be clear on what the communication strategy will be, including who the most appropriate spokesperson or 'media face' should be
During	• Liaise with other agencies involved in the response to ensure that messages are consistent, for instance on number of cases, where cases are occurring, what actions people should take and how the situation is changing
	• Provide regular updates that give honest and accurate information
	• Communicate regularly with other agencies to ensure limited resources can be allocated in the most efficient way
	• Provide regular briefings to the media and the public to ensure an information gap does not develop that can allow rumour and misinformation to spread
	• Monitor social media to understand public concerns and fears, and to pick up any incorrect rumours so that they can be countered quickly
After	• Ensure people who have been affected are aware of where they can get help and support in the long term if needed
	• Be aware that some people may not develop symptoms of emotional distress until some time after the event and ensure that they can access help when they realize they need it (see Chapter 9, this volume)
	• Instigate long-term health monitoring plans where needed in cases where health problems may develop long after the event (e.g. cancer screening programmes to a population exposed to radiation)

8.3.5 HOW: communications tools/methods

The start of an emergency is not the time to begin using new and unfamiliar communications systems: it is much better to continue using the systems and technology with which people are already familiar, although they may need to be used more often, or need more operators than usual to handle the additional capacity.

Fortunately, a health emergency is less likely to damage telecommunications infrastructure in the same way as a serious flood, a terrorist attack or an earthquake, so the usual lines of communication may be undamaged. But it is also the case that one of those other events may have triggered the health emergency, and so normal communications should not be assumed.

Planning in advance for what will happen if usual lines of communication are disrupted or unavailable, including what alternative methods of communication might be used, what to do if communications cannot be restored and what spontaneous communications networks might arise, will all help to manage communications during a real event. Guidance issued by the US Department of Homeland Security (see www.dhs.gov/emergency-communications-guidance-documents-and-publications) includes information on how to run a communications exercise (see Chapter 11, this volume).

In the modern world, the same message can be communicated by e-mail, post, text message, telephone, radio, television, Facebook, Twitter and many more. Messages will be sent and received by professional responders, members of the public, the media, politicians, employers and employees, community leaders, and friends and family. Each individual involved in a health emergency will receive (and probably send) multiple messages. With so many communication channels available, individuals are likely to first turn to the ones with which they are familiar, so it may be important to disseminate the message across a number of platforms to ensure that it reaches all demographics. The key issue, however, is to ensure that the message is consistent, as this will provide the greatest reassurance to those receiving it.

8.3.6 TO WHOM: the population at risk

The population at risk will benefit from having a clear and honest understanding of the risk, along with information on what they can do to manage it themselves, what aspects of it are being managed for them or where support for managing it is available. They will also need to know in what circumstances the risk may escalate or decrease.

Understanding trigger points for escalation at local and national level, and being warned of them in advance, will help with risk management and perception; for example, warning people in advance of when an emergency vaccination programme may begin, when local schools may need to close (and for how long) and how to call telephone helplines for further information. It is particularly important to consider how vulnerable populations can be reached. A particularly good handbook to follow is the Australian Government's *Communicating with People with Disability: National Guidelines for Emergency Managers* (see http://icrtourism.com.au/wp-content/uploads/2013/11/6_Communicating-with-People-with-Disability-National-Guidelines-for-Emergency-Managers.pdf).

In the UK, the Civil Contingencies Act 2004 places two distinct legal duties on Category 1 Responders (e.g. local authorities, blue light services and hospitals): (i) advising the public of risks *before* an emergency; and (ii) keeping it informed *during* an emergency. It also offers advice to emergency planners on how to develop communications arrangements that are appropriate for the message and the targeted audience.

8.4 Working with the Media

Media interest is an inevitable part of any emergency, and a health emergency will be no different. Working closely with the media, rather than seeing them as an enemy, can help to cascade public health messages to a wide audience and to ensure that public concerns are answered, rumours circulating on social media are countered and that regular updates can be disseminated to the public through televised briefings, for example. It will be important to consider and plan:

- Where are the media allowed (e.g. on hospital grounds; on wards; in public waiting rooms) and who should engage with them?
- How well is information about the emergency understood by media non-experts? Can this be improved so that better information is passed to the public?
- Hold regular press briefings and press conferences. If the media are handed information, there will be less need for them to go hunting for it.

8.5 Summary

Good communications strategies during a health emergency need to address communications to the media and the public as well as those between responder agencies. This chapter gives a broad overview of communications in relation to preparing for and responding to emergencies, and there are a number of good publications available on the Internet that have been produced to help people plan their strategies during generic emergencies and health emergencies in particular. Familiarizing oneself with these publications and taking the advice they give on planning, training and exercising will ensure that communications during a health emergency flow as smoothly as possible and do all that is practical to help ease pressure on the health care responders and bring the emergency to an end as quickly as possible.

Key Answers

- Why? Be aware of what might lead to a health emergency and put together as much information as possible in advance. Know how to receive and give out early warnings.
- Who? Plan your communications strategy well in advance of the health emergency. Know where information will come from and where it will need to be disseminated to. Know how to liaise with the media and how to use them as an ally.
- When? Plan for the different stages of an emergency. Different communication strategies may be needed at different times.
- What? Have a good idea of what information you will need to communicate. Know how to explain the situation in non-scientific language that can be easily understood by the public and the media.
- How? Know if, and how, communications during the health emergency will differ from everyday communications, and exercise any situation-specific systems and processes regularly. Know how communications can be scaled up if needed, including how communication systems will deal with any need for extra capacity. Be aware of spontaneous networks that may arise.

Further Reading

Australian Emergency Management Institute (2013) Communicating with People with Disability: National Guidelines for Emergency Managers. Available at: www.ag.gov.au/EmergencyManagement/Tools-and-resources/Publications/Documents/Handbook-series/handbook-5-communicating-with-people-with-disability.pdf (accessed 9 February 2016).

Brito, C.S., Luna, A.M. and Sanberg, E.L., for Police Executive Research Forum (2009) Communication and Public Health Emergencies: A Guide for Law Enforcement. Available at: www.bja.gov/Publications/PERf_Emer_comm.pdf (accessed 17 July 2015).

Cabinet Office (2012) *Chapter 7* (Communicating with the Public) of Emergency Preparedness, Revised Edition. Available at: www.gov.uk/government/uploads/system/uploads/attachment_data/file/61030/Chapter-7-Communicating-with-the-Public_18042012.pdf (accessed 17 July 2015).

Centers for Disease Control and Prevention (2015) Emergency Preparedness and Response: Crisis and Emergency Risk Communication *(CERC)*. Available at: http://emergency.cdc.gov/cerc/ (accessed 17 July 2015).

Cole, J. (2010) Interoperability in a Crisis 2: Human Factors and Organisational Processes. RUSI Occasional Paper. Available at: https://rusi.org/sites/default/files/201007_op_interoperability_in_a_crisis_ii.pdf (accessed 9 February 2016).

Cole, J. (2010) UK medical responses to terror threats. In: Richman, A., Shapira, S.C. and Sharan, Y. (eds) *Medical Responses to Terror Threats*. NATO Science for Peace and Security Series E: Human and Societal Dynamics, Volume 65. IOS Press, Amsterdam, pp. 63–74.

Cole, J. and Watkins, C. (2015) International employees concerns during serious disease outbreaks and the potential impact on business continuity: lessons identified from the 2014–15 West African Ebola outbreak. *Journal of Business Continuity and Emergency Management* 9, 149–162.

Department of Homeland Security (2015) Emergency Communications Guidance Documents and Publications: Operational Interoperability Guides. Available at: www.dhs.gov/emergency-communications-guidance-documents-and-publications (accessed 17 July 2015).

European Commission (2009) Strategy for Generic Preparedness Planning: Technical guidance on generic preparedness planning for public health emergencies. Available at: http://ec.europa.eu/health/preparedness_response/docs/gpp_technical_guidance_document_1_december_2009.pdf (accessed 17 July 2015).

Funk, S. and Jansen, V.A.A. (2013) The talk of the town: modelling the spread of information and changes in behaviour. In: Manfredi, P. and D'Onofrio, A. (eds) *Modelling the Interplay between Human Behaviour and the Spread of Infectious Diseases*. Springer, New York, pp. 93–102.

Gelfeld, B., Efron, S., Moore, M. and Blank, J. (2014) Mitigating the Impact of Ebola in Potential Hot Zones: A Proof-of-Concept Approach to Help Decisionmakers Prepare for High-Risk Scenarios Outside Guinea, Liberia and Sierra Leone. Available at: www.rand.org/content/dam/rand/pubs/perspectives/PE100/PE146/RAND_PE146.pdf (accessed 17 July 2015).

HealthC (2014) Improving Crisis Communication Skills in Health Emergency Management. Available at: http://healthc-project.eu/en/ (accessed 17 July 2015).

International Air Transport Association (2014) Crisis Communications and Social Media: A Best Practice Guide to Communicating in an Emergency. Available at: www.iata.org/publications/documents/social-media-crisis-guidelines.pdf (accessed 17 July 2015).

International Federation of Red Cross and Red Crescent Societies (2009) Communicating in Emergencies: Guidelines. Available at: www.eird.org/esp/ifrc-toolkit/guia/emergency-communications.pdf (accessed 17 July 2015).

World Health Organization (2008) Outbreak Communication Planning Guide. Available at: www.who.int/ihr/elibrary/WHOOutbreakCommsPlanngGuide.pdf (accessed 17 July 2015).

9 Psychosocial and Mental Health Care Before, During and After Emergencies, Disasters and Major Incidents

R. Williams[1] and V. Kemp[2]

[1]Emeritus Professor of Mental Health Strategy, Welsh Institute for Health and Social Care, University of South Wales, Pontypridd, UK
[2]Director, Healthplanning Ltd; and Associate, Welsh Institute for Health and Social Care, University of South Wales, Pontypridd, UK

Key Questions

- What do the terms 'psychosocial', 'psychosocial care' and 'mental health care' mean?
- Why is it important for emergency planners to understand the psychosocial and mental health needs of people affected by emergencies?
- What are the basic principles of psychological first aid?
- Are professional responders, including health care staff, less likely than members of the public to suffer the psychosocial and mental health impacts of emergencies?

9.1 Introduction

This chapter is intended as a basic briefing to assist health and social care organizations begin to take steps to prepare their staff and services for the psychosocial and mental health impacts of all kinds of emergencies, major incidents and disasters. This includes preparing for pandemic influenza, serious infectious illnesses, major terrorist events, and marauding terrorist firearms incidents. These plans, and the services required to deliver them, should be fully integrated into wider arrangements for emergency preparedness, resilience and response (EPRR).

This chapter provides a summary introduction to principles that are covered in more detail in Occasional Paper 94 (OP94) published by the Royal College of Psychiatrists in 2014 and other recent publications. This chapter offers an evidence-based framework that is consistent with policies of the North Atlantic Treaty Organization (NATO), countries in Britain and the Interagency Standing Committee's guidelines of 2008 for putting these principles into practice to meet the psychosocial and mental health needs of affected populations. Editorial policy limits the number of citations of sources within the body of the text. However, other authors' work is referenced at the end of this chapter.

9.2 The Psychosocial Approach

Understanding how people behave and their mental health needs before, during and after disasters and major incidents is of great importance when planning for disasters because it has implications for how:

- societies, communities and families plan and prepare for all kinds of disasters;
- responsible authorities provide public education and approach working in conjunction with communities to better understand and respond to their needs and preferences and ensure their continuing agency in large-scale disasters;
- governments and responsible authorities communicate with the public before, during and after events; and
- agencies manage events and respond in the immediate, short and medium terms.

Patel (2014) points out that there is a gap between the way in which mental health specialists apply the terms 'mental health' and 'mental disorder' and the broader conceptualizations of psychosocial suffering that affect very many more people than those who may require specialist mental health care.

The adjective 'psychosocial' refers to the psychological, social and physical experiences of people in the context of particular social, cultural and physical environments. It describes the psychological and social processes that occur within and between people and across groups of people. 'Mental health care' refers to delivering biomedical interventions from which people who have disorders may benefit. Usually, they also require psychosocial care as a platform on which their mental health care is based.

The number of people who require supporting interventions to assist them to cope with distress consequent on major incidents is very substantial, despite the majority of distressed people not being likely to develop a mental disorder. Many of them may be psychosocially resilient despite their distress. But, intervening early can reduce the risks of their developing disorders later. These interventions are termed 'psychosocial care'.

The approach recommended here espouses current professional opinion (Patel, 2014) that commends:

- distinguishing people who are distressed from those who require biomedical interventions;
- providing assistance for the greater number of distressed people through lower-intensity psychosocial interventions; and
- basing the distinctions between the two sorts of conditions on patterns and trajectories of people's experiences observed in general populations.

9.3 How People Behave After Emergencies: Myths and Misunderstandings

The most pervasive myth about disasters concerns panic. Persistence of this myth appears resistant to research evidence and, consequently, the frequency of panic is greatly exaggerated. Often, the term is applied, inappropriately, to describe goal-oriented, rational behaviour. In reality, panic occurs infrequently despite there being widely held beliefs that it is common after single-incident major events. Following Sheppard *et al.* (2006), it can be defined as having the following elements:

- behaviour that is intended to increase a person's chances of receiving apparently scarce or dwindling resources;
- putting personal safety ahead of assisting other people;
- 'contagiousness'; and
- irrational behaviours.

Panic, which meets these characteristics, is more likely to occur when: invisible agents are released during chemical, biological, radiation, nuclear or explosive/environmental (CBRNe) incidents; people feel powerless and/or are trapped; and people think there is no effective leadership or management that will prevent resources from being distributed unfairly on a first-come, first-served basis.

Some people who are affected by large-scale events that destroy the infrastructure may be immobilized by fear and helplessness and feel hopeless, although these responses are far from common. On the contrary, many people who are directly involved are first to take action; they are the first responders. There is evidence from many events of differing natures showing that many people are remarkably altruistic in the immediate aftermath; they behave in rational and selfless ways, even by putting themselves at greater risk to care for strangers. These findings have been described by research on the London bombings on 7 July 2005.

Other misunderstandings about how people and agencies respond to any major incident may be based on erroneous beliefs that:

- Everybody involved needs counselling or psychiatric treatment immediately after events.

- Post-traumatic stress disorder (PTSD) is the most likely psychopathology. PTSD, as defined by NHS Choices, is a disorder caused by very stressful, frightening or distressing events and can be caused by events such as serious accidents, sexual assault, mugging or robbery, and military service. While it certainly does occur after major incidents, adjustment and anxiety disorders, depression and substance misuse may be as frequent, or more frequent medium- and longer-term occurrences.
- First responders, health care workers, aid workers and military personnel are substantially unaffected by their work on emergencies.

It is also important to realize that there is little difference between the needs of first responders, including professional staff of emergency services, and the needs of the survivors whom they are seeking to assist. Everyone in the dynamic is at risk of psychosocial impacts and a proportion may require assessment for more specialized mental health care.

9.4 How People Behave After Emergencies: the Realities

There is a broad spectrum of ways in which people involved directly or indirectly in emergencies react emotionally, cognitively, socially, behaviourally and physically before, during and after events.

Everyone experiences stress in the course of their lives. One use of that term is to describe the challenges to people that may arise from untoward events that are of such a nature and/or severity as might cause them psychosocial motivation but also distress or harm. The people involved, including the staff of health care organizations, face primary and secondary stressors.

9.4.1 Primary and secondary stressors

Primary stressors are inherent in emergencies; they arise directly from the events. People who are affected are highly likely to suffer pain and distress. They and their relatives may undergo great upheavals and short-, medium- and long-term changes in their lifestyles as a consequence of their experiences, injuries, physical care, recovery and rehabilitation, and the effects on their families that continue beyond the injuries, illnesses and adversity they experience.

Lock *et al.* (2012) show that secondary stressors, by contrast, are circumstances, events or policies that are not inherent in events. Typically, the term describes conditions that persist for longer than the emergencies. It includes failure of infrastructure recovery, gaps in provision of services, failures in rebuilding and problems with insurance. Secondary stressors include the impacts of policies and plans made prior to events that limit people's recovery and adaptation, and sustain adversity.

Emotional experiences	Cognitive experiences
Shock and numbness	Impaired memory
Fear and anxiety	Impaired concentration
Helplessness and/or hopelessness	Confusion or disorientation
Fear of recurrence	Intrusive thoughts
Guilt	Dissociation of denial
Anger	Reduced confidence or self-esteem
Anhedonia	Hypervigilance
Social experiences	**Physical experiences**
Regression	Insomnia
Withdrawal	Hyperarousal
Irritability	Headaches
Interpersonal conflict	Somatic complaints
Avoidance	Reduced appetite
	Reduced energy

Fig. 9.1. Indicators of distress (reproduced from Alexander, 2005, with the permission of the Royal College of Psychiatrists Publications).

9.4.2 Indicators of distress in the immediate aftermath

Figure 9.1 summarizes the array of experiences that people may have in the immediate aftermath of disasters. Distressed people have a mix of some of these experiences.

Distress shortly after emergencies is very common. In most cases, it is transient and not associated with dysfunction. Distress that is more severe, more disrupting or associated with some limitations of function may be called 'acute stress' or 'post-traumatic stress'. The latter general term should be differentiated from PTSD, which, if and when it occurs, consists of a particular constellation of symptoms that tend to persist or worsen into the medium and long terms if sufferers are not treated. Before diagnosis, consideration should be given as to whether or not sufferers of persisting stress are experiencing secondary stressors that are sustaining their distress if it lasts more than several weeks or is more incapacitating. Also, some people may develop other mental disorders.

The majority of people affected by disasters are likely to benefit from lower-level, although none the less important, psychosocial interventions. Psychological first aid (PFA) is a good conceptual vehicle for initiating psychosocial care. Most people do not require access to specialist mental health care, but a substantial minority of people may do so and a small proportion of affected persons may require long-term mental health services in response to their needs. Therefore, a proportion of survivors who are thought to be at particular risk require surveillance and clinical assessment.

Box 9.1 summarizes the range of potential impacts of major incidents on people's psychosocial needs and mental health.

> **Box 9.1.** The psychosocial and mental health effects of disasters (reproduced from Appendix 1 in Williams *et al.*, 2014, with the permission of the Royal College of Psychiatrists Publications)
>
> **Direct effects on people who are affected**
>
> Primary and secondary stressors cause stress and, often, distress.
> 1. Immediate and short-term effects:
>   a. Resilient/non-disordered responses including short-term distress.
>   b. Acute stress reactions.
>   c. Neuropsychological changes in response to acute stress.
> 2. Medium- and longer-term effects:
>   a. Persisting distress maintained by secondary stressors.
>   b. Grief.
>   c. Mental disorders. (Note that these are very frequently co-morbid with other disorders) such as:
>   i. substance use disorders;
>   ii. adjustment disorders;
>   iii. PTSD;
>   iv. anxiety disorders; and
>   v. depression.
>   d. Impacts on personality
>
> **Direct effects of complex and sustained disasters on people who are at higher risk**
>
> 1. Distress (see text):
>   a. direct effects of complex multi-event disasters on people who are at higher risk; and
>   b. sustained distress that impacts on functioning.
> 2. Exacerbations of previous mental disorders of many kinds.
> 3. Onset of first episodes of common mental disorders.
>
> **Indirect effects**
>
> Disasters increase medium- and longer-term psychiatric and physical morbidity because they change the secondary stressors, medium- and long-term effects on social relationships, income and resources, and the societal conditions that shape mental and physical health through:
> 1. Increased poverty.
> 2. Changed social and societal relations.
> 3. Threats to human rights.
> 4. Domestic and community violence.

9.4.3 Psychosocial resilience

While the construct of resilience is not without critics, there is increasing evidence about psychosocial resilience. It describes how people, groups of people and communities may return to effective functioning, given adequate social support, after becoming distressed by emergencies and adversity.

Psychosocial resilience is not a synonym for resistance to the impact of events, absence of short-term distress after untoward events, or not suffering more prolonged distress if secondary stressors exacerbate it. Nor does psychosocial resilience describe absence of risk. It should not be inferred from positive mental health or absence of mental disorders. Neither should people who appear to be coping after major challenges be assumed to be unaffected psychosocially.

Instead, psychosocial resilience refers to people's abilities to adapt to, recover and learn from their experiences. Norris *et al.* (2009) defined psychosocial resilience as 'a process linking a set of adaptive capacities to a positive trajectory of functioning and adaptation after a disturbance'.

Most people cope well and recover given social support from relatives, friends and colleagues. Psychosocial resilience is an interactive, dynamic concept that describes interpersonal processes and the attributes of people by which they act singly and/or together to mitigate, moderate and recover from the effects of stressful events through exercising adaptive capabilities. Figure 9.2 portrays the three generations of resilience that have been identified.

The adaptive capacities that comprise psychosocial resilience are of genetic, psychological, social and environmental origins.

Box 9.2 lists our adaptation of the factors that Southwick and Charney (2012) have identified.

Many environmental factors affect how people cope with adversity including: their past and contemporary experiences and relationships; the severity and nature of their injuries and illnesses; access or otherwise to supportive family members and others; the health care and social resources available; the levels of adversity prior to events; and the adversity caused by events. Psychological factors include people's beliefs, attachment patterns, personalities, sense of agency, and tolerance of distress.

The quality of people's relationships either supports or undermines their resilience. Their abilities for forming and maintaining relationships with others at home and work as well as with strangers at times of greatest

First-generation resilience	Second-generation resilience	Third-generation resilience
The ability to cope reasonably well with events and their immediate aftermath	The ability to recover reasonably well from events	The ability of people to adapt in the light of lessons learned from events

Fig. 9.2. Three generations of psychosocial resilience (from Williams *et al.*, OP94, 2014, Royal College of Psychiatrists Publications).

Box 9.2. Resilience factors (adapted from the 10 factors of Southwick and Charney, 2012 with the permission of Cambridge University Press)

- Realistic optimism.
- Facing fear.
- Having strong guiding values.
- Spirituality.
- Social support.
- Physical fitness.
- Mental fitness.
- Cognitive and emotional flexibility and the ability to improvise.
- Creating meaning and purpose from events through personal growth.

need and accepting their support are key strengths. Social support consists of social interactions that provide actual assistance, but also embed people in a web of relationships that they perceive to be caring and readily available in times of need. These social relationships have very powerful influences on how we cope with adversity, ill health and emergencies. People who show good psychosocial resilience tend to perceive that they have, and actually receive, support. Leadership, good relationships between managers and employees, and peer support (Varker and Creamer, 2011) are recognized as playing vital roles in sustaining staff.

In summary, while psychosocial resilience is common, it should not be assumed and is not a reason for not taking action to provide interventions based on the principles outlined here.

9.4.4 Trajectories of people's responses

The trajectory of people's responses over time is an important feature of their reactions to threats or adversity. Several recent research studies show several patterns of how people respond over time (e.g. Bryant *et al.*, 2015). In broad terms, the patterns are:

- *Resilient responses* – depending on the nature of events, about 70% of people show psychosocial resilience. They suffer distress that is usually mild or moderate that rapidly reduces in severity if they receive support that they perceive as adequate.
- *Deteriorating responses* – some 10 to 20% of people may have symptoms and signs of stress that are initially of low severity, but which become more severe and/or associated with dysfunction over time. About half of this group may recover later while others develop chronic problems.
- *High initial stress responses* – about 10% of people may have high levels of stress before and immediately after events. The symptoms, signs and dysfunction suffered by about half of them may run a chronic course, while the course is an improving one for the others.

These generalizations suggest that the severity of stress that people experience in the period immediately before and after events and how they progress over time are good ways of separating people who will benefit from lower-level psychosocial interventions alone from those who also require specialist mental health assessments and treatments.

9.4.5 Risk factors

The differences in how people respond to disasters are influenced by their: personal characteristics; developmental and life experiences; training; family, team and group memberships; and the leadership and social support they are offered. People who are at greater increased risk of dysfunctional distress and social and mental health problems following disasters include: women; children and adolescents; older people; people who have pre-existing health problems and disorders; socially disadvantaged people; and staff of rescue and responding services.

Persons at greater risk of developing a mental disorder include people who:

- perceive they have experienced high threats to their lives or the lives of significant others;
- are physically injured;
- face circumstances of low controllability and predictability;
- live with the possibility that the disaster might recur;
- experience disproportionate distress or dissociation at the time;
- have experienced multiple losses of relatives, friends and close colleagues, and losses of property that is important to them;
- have been exposed to dead bodies and grotesque scenes;
- have endured higher degrees of community destruction;
- perceive that they have limited social support;
- are exposed to subsequent life stress;
- have been exposed to a major traumatic event previously; and
- have had a mental disorder previously.

9.5 Principles for Intervening

Effective approaches to EPRR and to planning and delivering the integrated emergency management cycle emphasize the importance of enhancing the capacities of people, families, communities and organizations for adapting well to new situations prior to events as well as afterwards, rather than solely recovering to the level of their abilities to function before events occurred.

9.5.1 General principles for early psychosocial intervention

The general principles for intervening psychosocially soon after events are:

1. Provide early intervention for everyone who is involved:

    - early interventions in communities after disasters should consist of social support and bolstering them rather than psychological treatments.

2. Provide practical interventions:

    - in proximity to where people are;
    - as soon as possible;
    - with the expectancy that people will recover; and
    - as simply as possible.

3. Base interventions on five core principles:
    - helping people to normalize their experiences while being aware that some people develop mental disorders;
    - enabling people by providing social support;
    - providing comfort through reflective listening and honest, accurate and timely information;
    - helping people to restore their agency and perceptions of themselves as effective persons; and
    - enabling people to seek further help.

4. Deliver support for people's social identities and receipt of social support through:
    - effective leadership;
    - restoring families and community groups;
    - re-opening schools;
    - restoring work opportunities; and
    - delivering psychosocial interventions based on the principles of PFA.

These principles apply equally to staff of the responding agencies who routinely require peer support, training and supervision.

9.5.2 Practical implications

There is a great deal that family members, colleagues and practitioners can do in the preparedness and response phases to alleviate people's suffering, accelerate their adaptation and recovery, and endeavour to prevent them from developing mental disorders in the medium and longer terms by offering social support and contributing to community development. OP94 (2014) indicates it is important that:

- All actions, interventions and service responses should promote a realistic sense of safety, calm, hope, empowerment, physical and social support, and access to welfare services.
- Efforts are made to identify the most appropriate sources of support (e.g. families, friends, communities, schools).
- Responses enable people who are involved to contact their families, and re-unite families as soon as possible.
- Service responses are based explicitly on people's human rights.
- Services facilitate appropriate communal, cultural, spiritual and religious healing practices. Memorial services and cultural rituals should be planned in conjunction with the people who have been affected.
- Local community leaders are involved in planning psychosocial and mental health care responses.
- Responders identify as early as possible those people who have more serious mental health problems.
- Events may affect people who do not live where incidents occur and so health care staff local to where people who are affected live (e.g. general

practitioners, practice nurses and staff of emergency departments) are made aware of possible psychosocial experiences and psychopathological sequelae.
- Responding organizations provide ease of access to specialist psychological and mental health assessments and intervention when required.
- Detailed planning occurs and authorities should be funded to deliver extra services to augment the existing primary and specialist mental health services for several years following disasters or major incidents.
- In the authors' opinion, certain specific formal interventions, such as single-session individual psychological debriefing for everyone affected, should not be provided.

9.5.3 Psychological first aid

The general principles predict core psychosocial interventions in the immediate aftermath that are now gathered together within a construct termed PFA. PFA is not a therapeutic intervention but its components are intended to reduce people's initial distress in the immediate aftermath of disasters and foster adaptive functioning by encouraging people to influence and adapt to the sources of stress. It consists of activities to: provide affected people with effective communications; ensure that they are offered social support; restore people's agency; reconnect people with family members; and begin to restore communities. PFA underpins all levels of care and is also at the core of approaches required by staff of responding agencies.

9.6 Caring for Responders and People Who Intervene

The extent and frequency of the psychosocial and mental health impacts of emergencies and disasters on rescuers, the staff of emergency services, and health and social care agencies fall between epidemiological assessments of the impacts on people who are directly involved and people who are not involved.

Evidence about what should be done to provide training and support for people who intervene is available (Williams and Greenberg, 2014). OP94 contains checklists of important actions that leaders and staff should take to prepare responding staff before they become involved, support them while they are working in a disaster area and assist them to recover afterwards. Core to supporting people who respond and intervene are the vital necessities of ensuring that they are well briefed, well led, and offered sufficient social and peer support.

9.7 The Implications for Responding Agencies and Practitioners

This chapter shows that there are important matters relating to psychosocial and mental health care that should be considered before and during each of the main phases for responding to disasters. Knowledge of how people cope and the importance of their relationships, social support, leadership and care should inform all EPRR plans, including actions to support and develop communities before disasters occur and to intensify and maintain community development alongside personal psychosocial care in the response and recovery phases. Additionally, developing third-generation psychosocial resilience is a vital consideration in the mitigation phase for reducing risks and threats. This agenda gave rise to the strategic, stepped model of care that is core to NATO's guidance.

The approach recommended is to provide psychosocial care, otherwise termed 'public mental health interventions', that are intended to prepare people before disasters strike, and then support their coping in the immediate aftermath of events and through their medium-term recovery. A number of early interventions lack evidence of preventive effect. However, a number of studies show that high levels of perception of social support are protective and associated with lower rates of PTSD. So practical and pragmatic support is also oriented to endeavouring to prevent people from developing serious mental disorders (Bisson, 2014).

Despite best endeavours, a minority of people are at risk of developing new mental disorders and more may experience exacerbations of previous disorders. People in these groups require timely personal mental health care.

In summary, there are three objectives in the short and medium terms:

- making psychosocial care available to everyone who is affected;
- prevention of suffering and, if possible, people developing mental disorders; and
- providing services that offer surveillance for people who are more at risk of developing mental disorders and mental health care for people who require it.

These objectives require a model of care that can be incorporated into EPRR plans.

9.7.1 A strategic, stepped model of care

The strategic, stepped model of care is a basis for planning responses to meet people's psychosocial and mental health needs and is intended as a practical, conceptual resource for planners. It links the impacts of events with core components of psychosocial care and mental health care that populations of people, communities and particular people require. It balances the need to blend public health approaches for populations of people with personalized health and social care interventions that people at greater risk may require. It has seven cumulative steps that fall into four groups.

Strategic and operational preparedness
1. *Strategic planning*: comprehensive multi-agency planning, preparation, training and rehearsal of the full range of service responses are required in advance of events.

2. *Prevention services*: prevention services that are intended to develop the collective psychosocial resilience of communities and families should be planned and delivered in advance of events.

Public psychosocial care after events

3. *Psychological first aid*: families, peers, communities and trained laypersons should be supported in providing responses to people's psychosocial needs in the aftermath that are based on the principles that underpin the construct and components of PFA.

4. *Community support and development*: community interventions include providing activities that are intended to sustain communities, restore their cohesion and develop their abilities to deliver social support through, for example, leadership and action to restore access to social encounters, more ordinary daily routines and future-oriented learning in schools and communities. Families and communities should be provided with routes by which they can draw the attention of the personal health and social care services to people in need of them.

Surveillance and other public mental health interventions

5. *Identifying people who have unmet psychosocial and mental health needs*: services should be established that are able to initiate public mental health assessments that are designed to identify, after the immediate aftermath of events, unmet needs of people whose distress is sustained by secondary stressors and people who need personal assessments in case they have emerging or recurrent mental disorders.

Personalized psychosocial and mental health care

6. *Access to primary mental health care and social care services*: primary health and social care services should be augmented in the aftermath and medium term after disasters to provide surveillance, and assessment and intervention services for people who do not recover from immediate and short-term distress or develop problems later. Research on the effects of the earthquake sequence of 2010–11 in Christchurch, New Zealand substantiates this advice and the importance of supporting staff of the primary health and social care services (Chapter 17, this volume).

7. *Access to secondary and tertiary mental health care services*: planners, health care strategists and commissioners should ensure that there are routes that are negotiated and well promulgated to and through primary health care, social care and education services for people who appear to require access to specialist (secondary and tertiary) mental health care. The arrangements made should enable people to be referred with a minimum of delay.

This model of care is summarized in Fig. 9.3.

9.7.2 The roles of specialist mental health practitioners

It is important that the specialist mental health services should focus their activities on people who might have mental disorders that require specialist intervention (Step 7).

Strategic intent	Purpose of activity	Phased objectives	Action steps	Timescale
Developing and sustaining the collective and personal psychosocial resilience of communities, families and staff groups	Preparedness, risk communication and mitigation	Preparedness	1. Strategic planning	Continuing
		Mitigation	2. Develop community resilience	
	Delivering public welfare, social and health care paradigms of response to psychosocial needs	Supporting people	3. Psychological first aid	Immediate and continuing
		Supporting and sustaining communities	4. Community development	
Making transition from public health to personal health care responses to people's needs	Making available initial public mental health assessments to identify the needs of people whose distress is sustained by secondary stressors and people who need personal assessment	Identifying people who have unmet psychosocial and mental health needs	5. Surveillance of people at risk and direct signposting of people in need to the most appropriate health and social care services	After 2–4 weeks and continuing into the medium and longer terms
Delivering responses to people's personal psychosocial and health care needs	Delivering personal psychosocial and health care paradigms of response to people's needs for mental health care	Assessing and triaging people who have been identified as possibly requiring mental health care and offering primary mental health care	6. Augmented primary health and social care	Medium and longer terms
		Delivering specialist mental health care for people who have been assessed as needing it	7. Specialist mental health care	Medium and long terms

Fig. 9.3. The strategic seven-step model of care (© Richard Williams & Verity Kemp, 2015, reproduced with the permission of the copyright holders).

However, it is important that staff of the specialist mental health services and social care practitioners should also offer:

- strategic and operational advice to the responsible authorities before, during and after events (Steps 1 and 2);
- advice to communities and supervision of laypersons (Steps 3 and 4);
- advice to the responsible authorities about their meeting the requirements for surveillance (Step 5); and
- to augment the services delivered by primary care (Step 6).

9.7.3 Specific components of the responding services

This model of care underpins the principles covered by OP94, which provides information about the specific components of the responding services that are required within:

- the first week of a disaster or major incident;
- the first month of a disaster or major incident;
- 1–3 months after a disaster or major incident; and
- beyond 3 months after a disaster or major incident.

OP94 is freely available on the Royal College of Psychiatrists' website (www.rcpsych.ac.uk/usefulresources/publications/collegereports/op/op94.aspx). Readers may refer to it and the References and Further Reading below for detail that is beyond the scope of this introductory chapter.

Key Answers

- 'Psychosocial' refers to people's emotional, cognitive, social and physical experiences in the context of particular social, cultural and physical environments. It describes psychological and social processes that occur within, between people and across groups of people. 'Psychosocial care' describes interventions that offer comfort and relief from suffering, improve how people cope with adverse events, and also endeavour to reduce the risk factors for people developing psychiatric disorders. 'Mental health care' refers to delivering bio-social-medical interventions from which people who have disorders may benefit. Usually, they also require psychosocial care as a platform on which their mental health care is based.
- Distinguishing how people actually behave and their mental health needs after emergencies, rather than relying on inaccurate myths, is important when planning for disasters because this has implications for how best to prepare communities and assist them to develop their collective resilience, and reduce the disruption and the psychosocial consequences that arise. This knowledge also enables people to develop personal resilience and may contribute to reducing the risk factors for later psychiatric disorders. It enables agencies to respond proportionately, flexibly and in timely ways to the needs and preferences of people who are affected that include an integrated continuum of care and recognizes that people's needs are immediate and may continue in the medium and long terms, while taking account of the needs of responders and staff of services.

Continued

Key Answers Continued

- The practices that constitute PFA seek to reduce distress and ensure that people's basic needs are met following an emergency or major incident. It is based on the principles of achieving safety, calm, connectedness, hope and self-efficacy. It aims to provide early assistance and is ideally initiated by people first on site who undertake an initial assessment of the psychosocial impacts of the event. Using PFA: supports stabilization and reduces further psychosocial problems; helps to maintain psychosocial status until professional mental health care is available, if required; and promotes faster and better recovery.
- Work in rescuing people from major incidents, with survivors to meet their psychosocial and mental health needs, and with people who have suffered abuse or recurrent traumatic events or relationships is challenging and stressful. There are stressors for all responders inherent in their work, including exposure to gruesome sights, smells and events, on-site dangers and interpersonal violence, and exposure to survivors' stories. This work can provoke feelings of powerlessness caused by responders' feelings of limitations with providing help at the level needed. Events in which people die, children are involved, and when violence is directed towards uninvolved people tend to produce most distress for responders.

References

Alexander, D.A. (2005) Early mental health intervention after disasters. *Advances in Psychiatric Treatment* 1, 12–18.

Bisson, J.I. (2014) Early responding to traumatic events. *British Journal of Psychiatry* 204, 329–330.

Bryant, R.A., Nickerson, A., Creamer, M., O'Donnell, M., Forbes, D., Galatzer-Levy, I., et al. (2015) Trajectory of post-traumatic stress following traumatic injury: 6-year follow-up. *British Journal of Psychiatry* 206, 417–423.

Lock, S., Rubin, G.J., Murray, V., Rogers, M.B., Amlôt, R. and Williams, R. (2012) Secondary stressors and extreme events and disasters: a systematic review of primary research from 2010–2011. *PLoS Currents Oct* 29, 4 (doi:10.1371/currents.dis.a9b76fed1b2dd5c5bfcfc13c87a2f24f).

Norris, F.H., Tracy, M. and Galea, S. (2009) Looking for resilience: understanding the longitudinal trajectories of responses to stress. *Social Science & Medicine* 68, 2190–2198.

Patel, V. (2014) Rethinking mental health care: bridging the credibility gap. *Intervention* 12, 15–20.

Sheppard, B., Rubin, G.J., Wardman, J.K. and Wessely, S. (2006) Terrorism and dispelling the myth of a panic prone public. *Journal of Public Health Policy* 27, 219–245.

Southwick, S.M. and Charney, D.S. (2012) *Resilience: The Science of Mastering Life's Greatest Challenges.* Cambridge University Press, Cambridge, UK.

Varker, T. and Creamer, M. (2011) *Development of Guidelines on Peer Support using the Delphi Methodology.* Australian Centre for Posttraumatic Mental Health, University of Melbourne, Melbourne, Australia.

Williams, R. and Greenberg, N. (2014) Psychosocial and mental health care for the deployed staff of rescue, professional first response and aid agencies, NGOs and military organisations. In: Ryan, J., Hopperus Buma, A., Beadling, C., Mozumder, A. and Nott, D.M. (eds) *Conflict and Catastrophe Medicine: A Practical Guide.* Springer, London, pp. 395–432.

Williams, R., Bisson, J. and Kemp, V. (2014) *OP94. Principles for Responding to People's Psychosocial and Mental Health Needs after Disasters.* Royal College of Psychiatrists, London.

Further Reading

Alexander, D.A. and Klein, S. (2009) First responders after disasters: a review of stress reactions, at-risk, vulnerability, and resilience factors. *Prehospital and Disaster Medicine* 24, 87–94.

Hobfoll, S.E., Watson, P., Bell, C.C., Bryant, R.A., Brymer, M.J., Friedman, M.J., *et al.* (2007) Five essential elements of immediate and mid-term mass trauma intervention: empirical evidence. *Psychiatry* 70, 283–315.

North Atlantic Treaty Organization (2009) *Annex 1 to EAPC(JMC)N(2008)0038 Psychosocial Care for People Affected by Disasters and Major Incidents: A Model for Designing, Delivering and Managing Psychosocial Services for People Involved in Major Incidents, Conflict, Disasters and Terrorism.* NATO, Brussels.

Williams, R., Kemp, V.J. and Alexander, D.A. (2014) The psychosocial and mental health of people who are affected by conflict, catastrophes, terrorism, adversity and displacement. In: Ryan, J.M., Hopperus Buma, A.P.C.C., Beadling, C.W., Mozumder, A., Nott, D.M., Rich, N.M., *et al.* (eds) *Conflict and Catastrophe Medicine: A Practical Guide*, 3rd edn. Springer, London, pp. 805–849.

World Health Organization, War Trauma Foundation and World Vision International (2011) *Psychological First Aid: Guide for Field Workers.* World Health Organization, Geneva, Switzerland.

10 Business Continuity

J. Hebdon

Emergency Preparedness, Resilience and Response Senior Officer, NHS England Central Team, UK

> **Key Questions**
> - Why is business continuity so important for all organizations?
> - What is the PDCA cycle?
> - How do you develop a business continuity system?
> - How do you assess the impact of an incident on your organization?

10.1 Introduction

Business continuity is important for all organizations, from single-handed general practitioner practices to multisite trusts. It is not only important to protect an organization's reputation and the safety of staff, patients and customers, but it also gives the organization an underpinning understanding of its functions and processes and their importance. This understanding will enable easier creation of emergency plans, especially those that rely on the collapse of other functions to protect or provide a new service. It should not be underestimated how much business continuity planning will help other preparedness areas; for example, business continuity workforce planning gives a framework to support pandemic influenza planning, winter resilience plans and planning for industrial action. It is also important not to forget that the functions in the emergency plans will need to be protected by contingency arrangements of their own; for example, how decontamination will occur without a decontamination facility.

10.2 Alignment or Accreditation?

It is worth noting that NHS organizations in the UK are required within the Emergency Preparedness, Resilience and Response Core Standards to align to the international standard for business continuity management systems (ISO 22301). The author has taken this to mean that the business continuity management system (BCMS) should meet the requirements or look to meet them, but without paying for the formal auditing and accreditation by the standards agencies. This approach allows some flexibility and means it is possible to interpret the standards and devise ways to meet them. It also allows consideration of the cost implication of accreditation; unless there is a specific requirement of a supplier or under law, it may be seen as excessive to spend money on this.

10.3 Development of Business Continuity Systems

With business continuity it is possible to start from anywhere in the *plan, do, check, act* (PDCA) cycle (Fig. 10.1) as indicated by the international standard in business continuity. The advantage of this is you can choose the most appropriate starting point for your organization; for example, if a system is already established it can be methodically reviewed within the 'check' part of the cycle to ensure that changes to the system are appropriate. Many people will want to do this when they first join an organization; a new starter won't make friends with the people using the system if he/she decides to scrap previous work and start again without justification.

The following sections describe the components of the PDCA cycle and what needs to be considered. It should be stressed the key to a good

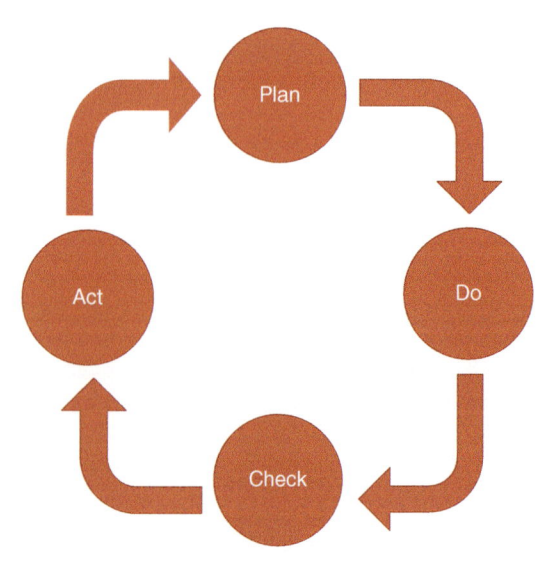

Fig. 10.1. The PDCA cycle.

system is ensuring that the information needed at each stage to inform the next stage is produced keeping things as straightforward as possible, so people can understand the system without too much trouble and still give the information required.

10.4 Plan

Assuming a start from nothing, then planning and developing the system will be the first step in getting a BCMS established in the organization. It is important to plan how things will work together to provide the simplest way to get information. Before starting the development of the system, ensure an understanding of which other management systems are in place, how developed they are and who owns them.

A good example is risk management. This will give a process by which risk should be assessed and managed, and the acceptable levels of risk for use in the BCMS. This negates the need to devise a separate process, which could lead to issues with auditing the risk management system. Another quick win is to use established document control processes to support the BCMS being created, as this will have recognizable elements to those in the organization. This is not to say these processes cannot be tweaked or built upon within the BCMS, but make it clear that these apply only in addition to the processes in place and do not remove elements of these processes if all documents within the organization have to meet them.

Once this background information has been reviewed there will be an understanding of how to approach some of the key elements of the BCMS, and there may be specific policy templates that must be followed when writing and completing the new policy documents for the BCMS. It is important to include the key policy requirements from ISO 22301 as this will allow later demonstration of alignment, if chosen by the organization.

Policy development should be focused on the intentions of the organization to carry out a BCMS and establish why the organization is placing importance on this, the context of the system, define some limitations and be signed off at the highest level of the organization. It may be easer here to split out *how* the policy will be achieved; this is especially important if the document is to be reviewed by a committee with no understanding of the processes involved. That isn't to say that the statement of policy intention cannot sit with how the policy will be implemented. It may be useful to combine the business continuity policy with wider emergency preparedness information. This can bring the intention of the organization to meet its requirements under legislation and guidance together under a single document, thus providing a basis for the areas to coexist and work together for common purpose.

Having the organization committed to the development of the BCMS it is time to think about how this will be done, the actual system to manage the plans. A document or series of processes need to be compiled to ensure the BCMS is understood by the ends users. Box 10.1 outlines what this needs to consider.

> **Box 10.1.** Business continuity documentation outline
>
> - How the aim will be achieved.
> - What will be done with information from the business impact analysis.
> - How training will be conducted and its success evaluated.
> - Document controls.
> - Risk assessment.
> - Standards for plans.
> - The review and auditing processes to be used.

It is often effective to write all aspects down as it allows anyone auditing the BCMS to see and understand the processes being used and to monitor compliance. It may also be useful to establish some key performance indicators (KPIs) in some of the sections to monitor progress. These need to have realistic and achievable targets if the system is to be seen as effective and adding worth to the organization (Fig. 10.2).

The organizational risk assessment is a key component of the BCMS that needs to be undertaken to identify potential business continuity risks. It is important to link with the existing organizational risk management system to prevent duplication and, where practical, to use standard assessment formats. For the NHS, this is produced by the NHS Litigation Authority; the US Department of Defense also uses a standard assessment matrix; and other government departments may have a standardized approach such as the UK Home Office matrix for National Risk Assessment. Identified hazards may be grouped into some generic clusters, such as staff loss, premises disruption, data and data systems, and suppliers and contractors. This enables the production of single, multi-layered strategies that will protect against disruption in these general areas and reduces the need to plan against each hazard individually; see Case Study 10.1 for an example of how this may be achieved.

10.4.1 Business impact assessment

Business impact assessment or analysis (BIA) is a major component of the BCMS. Completing the BIA will allow the organization to know resource levels for its services, the impact of disruption and how important each service is within the organization. Identifying the latter will be important in deciding how much resource to aim at recovering services during a disruption. It will also be important to work out how each service relates to others within and outside the organization, again allowing a better identification of the importance of each service. A key aspect of this is identifying the management services that support the organization. A good example is payroll: all services are dependent on it and all services provide it with information; however, if the payroll is disrupted you will have lots of annoyed staff and a big reputational impact. The other aspect to consider is that when assessing impact, the focus should not be on identifying how likely the event is but the impact of it occurring.

Plan compliance	%
Number of staff trained	%
Plans exercised	Yes/No DATE

Fig. 10.2. Business continuity key performance indicators.

Case Study 10.1 Business continuity and staffing

There may be hazards to staffing caused by various events such as pandemic illness, epidemic illness, transport strike, a lottery syndicate win or a group of people tendering their resignation at the same time, which will all reduce the level of available staff from a full service complement. Therefore, it is best to develop a response strategy that can manage the impact of all of these to the services or organization. Assuming the range of staffing loss will be manageable, there will be a series of different options to enact to allow the service to continue. It is always important therefore to produce multiple strategies and look at how resilient these are. Some of these strategies will be in place at all times for the organization to manage such as the sickness absence policy, which aims to balance the lack of staff with suitable return periods following illness, etc. Striking the right balance will be important not only to reduce absence, but also ensure that more people are not infected by staff returning too soon.

When designing the BIA, it is important to ensure information that is not going to be used or is already available is not requested. This is important in order not to discourage those working with the system; especially if the size and complexity of the request being made could be considered excessive. People contributing to a BIA must be able to understand the worth of the questions being asked in order to ensure they give good-quality answers.

10.4.2 Plan development

The next part of the BCMS is the development of a business continuity plan or plans to respond to the disruption. It is recommended that a template is used which sets out specific areas of consideration, communication routes, has version controls and a clear scope. Consideration should be given to the level of operation of these plans to ensure they contain the correct elements. An overarching document that gives the response to actual incidents and the coordination of response and recovery across various services through the organization will be needed to draw these different levels together. Below this plan there may be service-level plans, specific to areas or teams of the organization, that are based on the information in the BIA. In lower-level plans there will be a focus on the functions offered, how important these are to the service plans to deal with incidents within the service and local escalation routes. These standards should be set out in the BCMS document being created and the plans must reflect this through the template. Reviewing this over time through engagement with plan owners will need to be considered too.

Finally, it is important to consider how the organization will interact with key stakeholders, including those that provide services or supplies to the organization. These should be captured during the BIA process, but a process will be needed to decide how assurance will be sought from these organizations about their own business continuity arrangements and if they are sufficient to keep meeting the contracting organization's needs. It is also important to understand how important the organization's business is to these stakeholders, and what they need to be told during an incident. This is especially important in networks and system flows where impacts in one place ripple through to other areas and steps are needed to ensure that these are minimized. The classic example of this is in winter where mortuary demand rises and the number of internments decreases, creating pressure in one part of the death management system. In this situation capacity has to be managed but the other parts of the death management system need to understand the pressure on mortuary capacity to recover quickly, working together to remove the pressure and prevent failures, excessive costs and reputational damage. One issue here is transferring the risk out of the organization but not the cost: the organization may still need to provide the service itself in the event of the contractor or supplier actually failing, so this must be included in the plans being developed.

10.5 Do

Implementation of the BCMS is often very complicated as it involves winning over people who will not see business continuity as a priority in their own list of priorities. Therefore, time must be given to address how the process will assist individuals and their services. A focus on the BIA outcomes can also trigger anxiety among staff, who may see the process as one looking at what could be cut out from the business.

Keeping these things in mind, it is important to sell the benefits and legal protections that planning give and focus on these, while introducing people to the processes they need to follow. In any process, users will need to be empowered to complete assessments and plans, the skills and basic knowledge of what is wanted and how it is going to be used. It is important to recognize that no one individual will have all the answers and support will be needed from a specialist if the right information in the right format in the right box is going to be achieved.

Medical services frequently run with a lead manager, a lead nurse and a lead medic, therefore they need to all have time to input into the planning and BIA to get the best representation of impacts and also the best response in the event of a disruption. This will also reduce the assumptions made about how or what each different group is capable of when planning the response and recovery from a disruptive event.

There is a need to focus planning at the right level (organizational or service level) as both will have the power to influence outcomes and strategies for responding will need to be developed across these. For example, a service needs to have a plan for how it will respond to no electricity while the estates and facilities team investigates and implements its own plans to restore the supply. The focus of the response will be different for these groups: the service to reduce harm to its users, while the estates and facilities team will be focused on restoring the electricity. Services need to be encouraged to plan based upon their own priority, as a lower-priority service may need to operate without a resource for longer than a higher-priority one.

10.5.1 Training

Training will be a substantial part of any BCMS and is split into two areas: (i) training in use of the BCMS; and (ii) training in incident response. The latter needs to be led locally for each service's own response and centrally for response teams that will coordinate between services and leaders.

Training in the use of the BCMS will need to be focused on what needs to be done and how in the planning process to get the correct information. The person delivering the training should be the business continuity system owner and the training focused on the most basic level of how to run the individual parts of the BCMS. Training needs to be focused according to the different roles within the BCMS so the correct people understand what they need to do to make the system work. This runs throughout the organization, as all staff need to have a basic understanding that there is a system and how they contribute to it.

The training in incident response will need to be focused according to the roles of individual services and on the localized response. It should consider what alternative working looks like and how and who implements it. This is really the practical side of being able to respond. Time and again business continuity plans fail when IT systems go down and staff are unable to use the manual processes because they have not been adequately trained in them or they are unsure if they should be using them or not.

Centrally, those required to coordinate the response across the organization need to understand their own role within plans and how they will collate large volumes of information across the organization. This is especially important when gaining information on staffing levels or other resources needed in a response and if the movement of resources is needed to maintain a basic service provision. Leaders in the response need to understand if they have specific duties within the plans, such as communicating or designating responding staff, and it must be clear to those who might respond how they fit into response structures. This is important if strategic decisions are to be taken only by specific persons in the response.

An important thing to remember about training is that it should be influenced by those being trained. Assessing how those trained feel about carrying out tasks and their own preference for further training will assist in the development of future training content. This will allow those who are undertaking tasks to focus on their own needs while allowing the system and response to operate. Evaluating the training will also give information that can be used as a KPI in the measurement of system success.

10.6 Check

Once the system has been implemented it is necessary to ensure that the responses that are planned will work and the training has been effective. This can be achieved through testing and exercising of the component parts of the business continuity response strategies. Exercises are used to identify gaps in the strategies, areas that need to be added, check adequate resources and ensure that the training undertaken is sufficient to allow those needing it to carry out their response tasks. It is important that those participating in exercises understand there are no pass or fail criteria and that the process leading to decisions is what is being exercised. Exercises often bring up specific response issues and it is important to remember this when developing corrective actions for the BCMS, otherwise it may skew the response towards a specific scenario that is unlikely to occur in the way it was exercised.

Testing is usually done on component parts of the business continuity strategies that can either succeed or not, and the system should be set up to identify these and ensure they are captured. Testing may focus on the IT systems and automated backup systems such as generators and disaster recovery tapes, where the hardware is either working or it is not. Of course, having a working generator on any given day does not mean the organization can rule out planning for the loss of electrical supply because a fault may occur that does not trigger this line of defence. In all cases services should plan to not have a resource for a period of time. Regular testing of the different components that support business continuity strategies will allow a better response, an earlier identification of issues and lead to problems being spotted before an incident calls the equipment into use.

Testing routines are important to capture learning, but the results of these tests must be handled in an appropriate manner. In the author's own experience, a system was in place to test the IT backup tapes but the success of these tests not monitored for a considerable period of time, which left a section of data compromised on a server during a severe failure. Audits can be conducted in various ways and on different parts of the BCMS as required; two key components will be internal audit on the system and external audit conducted independently of the organization.

10.7 Act

The final part of the PDCA cycle is to act. This ensures the organization takes actions to correct problems with compliance within the system and to make changes to the system itself. Using the results of exercises, tests, audits (external and internal) and getting general feedback from system users will allow problems across the system to be identified. It is important not to become overly focused on one area. Exercises and tests will allow improvements to be made to the response strategies, while audits and user feedback will allow targeted improvements to be made to the system that has been implemented.

Exercise and test results should be reviewed for how they impact the response in different scenarios. It is important to develop a generic response, rather than making an improvement to the system that will skew the response in a different scenario. If something is specific to the scenario exercised, consider how this can be implemented without creating specific documentation and duplicating effort. Performing root cause analysis before deciding on the actions to take is an important part of what has happened. There may be several actions to take ranging from training improvements to action prompts. In some cases, there might just be a need to present information in a more accessible way – this often is the case where no diagrams of a process exist, only blocks of text that are difficult to follow.

Management of these actions will be important to ensure continued development and show that the system is open to change and the necessary steps are being taken to address deficiencies. This will be critical for an organization wishing to avoid potential litigation at a later date. Identifying an issue but failing to address it could easily result in a successful prosecution where the change would have prevented loss of life or serious injury. This may even extend to loss of data if operating under a law that requires the protection of any data being processed by the organization. A simple summary document of all the actions is a straightforward way of doing this; each action can be coded to allow identification of where it came from, monitor any actions that need to be closed urgently or are well overdue, and what is preventing these being closed.

The organization's management needs to approve the actions where multiple options exist, so the system that has been created must address the level at which this must occur and who has authority at each of the different levels. Minor amendments are often suitable to be fixed by the system or strategy owner. Anything with cost or resource implications should be highlighted in the management review documentation and approved at the most senior level.

Creating documentation for the management review should be straightforward for a well-established system where there are KPIs, a clear structure and a series of coordinated actions. The review should include a current status of all these indicators, a summary of the actions open and closed, and key decisions that are needed by top management in order for the system to continue to function correctly.

As a minimum this review should be annual; however, it might be suitable and allow greater transparency in the system if a short summary is provided every 6 months. This will depend on review and exercising schedules of plans. The key is to keep any review digestible and easily understood, as the system will benefit if issues with compliance and good practice are understood at this top level.

There will be a requirement to look at changes to legislation and recommended practice and include these within the review to ensure the organization remains within the law and has an effective system. All decisions by this group on the BCMS should be recorded and added to any action log that is being kept to manage the system. The natural conclusion to any of the reviews will be to return to the planning of the system to ensure the changes approved can be incorporated fully throughout the system as required.

The cycle chosen for the organization will most likely depend on the maturity of the system and the frequency of change – twice per year is a manageable period initially when establishing the system, but more frequently than this and the system does not have time to become embedded. Consideration needs to also be given to the understanding and sympathy afforded by those implementing the system if it undergoes change too frequently.

> **Key Answers**
>
> - Business continuity is important for all organizations, particularly so for those that have a responsibility to provide services in an emergency. Business continuity not only helps protect an organization's reputation, the safety of its staff, patients and customers, but also gives the organization an underpinning understanding of its functions and processes and their importance.
> - The PDCA cycle is a framework to support the development of business continuity arrangements. It stands for plan, do, check and act. Planning provides the framework to describe the actions to be taken during an incident to preserve critical functions. The next stage of 'do' helps embed these arrangements into the teams and organization. 'Check' ensures plans and arrangements are adequate and fit for purpose. Finally, 'act' takes the learning from exercising and testing and ensures it is fed into the plans and arrangements so that they become more evidence-based.
> - At the start of the business continuity management (BCM) process it is important to develop and set out the business continuity system. This important step sets out how to plan for things to work together to get information in the simplest way. The initial actions will include the development of a BCM policy setting out the organization's intentions, links to the organization's risk register and the development of a business impact analysis (BIA).
> - The impact of an incident on an organization can be assessed through undertaking a BIA, which identifies the resource levels for services, the impact of disruption and how important each service is within the organization. It can also help identify the potential impact of a scenario on a range of metrics including financial, staffing and reputation.

Further Reading

Business Continuity Institute (2013) BCI Good Practice Guidelines. Available at: www.thebci.org/index.php/resources/the-good-practice-guidelines (accessed 18 December 2015).

British Standards Institution (2012) ISO 22301 Business Continuity Management. Available at: www.bsigroup.com/en-GB/iso-22301-business-continuity/ (accessed 18 December 2015).

International Organization for Standardization (2012) ISO 22313:2012: Societal security – Business continuity management systems – Guidance. Available at: www.iso.org/iso/catalogue_detail?csnumber=50050 (accessed 18 December 2015).

Sterling, S., Duddridge, B., Elliott, A., Conway, M. and Payne, A. (2012) *Business Continuity for Dummies*. Wiley, Chichester, UK.

11 Training and Exercising for Emergency Preparedness, Resilience and Response

R. Ellett[1] and A. Wapling[2]

[1]*Former Fire Commander, Humberside Fire and Rescue Service; and Course Director at the Emergency Planning College, Easingwold, Yorkshire, UK*
[2]*Regional Head of Emergency Preparedness, Resilience and Response, NHS England (South), UK*

Key Questions

- Why is it important to train staff to manage emergencies?
- How can specialist training best be delivered to those who need it?
- How do we evaluate training?
- Why is it useful to test our preparedness through exercises?
- How does learning make our response to emergencies more effective?

11.1 Introduction

Emergencies are a unique and sometimes extreme environment not only for those caught up in the event, but also those responsible for enacting the response. For emergency responders it is important to recognize that the environment will be different from that of their daily work. The response of an ambulance paramedic to a road traffic collision with two cars (a relatively frequent if unfortunate event) will be different from that of a collision of a coach with 52 children involved (thankfully a rarer occurrence). As emergency planners and organizational managers we have a responsibility to prepare and equip our staff to manage extreme and uncommon events.

This chapter considers the training and exercising that is required to support responders in gaining the skills to manage emergencies. The first half focuses on specific training elements required to deliver successful emergency responses, while the second considers the possibilities to test and exercise the training.

11.2 Training

When identifying training needs and solutions, emergency planners are encouraged to examine non-health related sources. With *isomorphic* meaning 'being of identical or similar form, shape or structure', the phrase *isomorphic learning* is used to refer to how lessons can be identified and translated from one situation into another: in this case, taking the experiences and lessons identified in other disasters and applying them to one's own type of organization (Toft and Reynolds, *Learning from Disasters*, 3rd edn, 2006).

11.2.1 The training challenge

In terms of mandatory training in the UK for health emergency preparedness, resilience and response (EPRR), there is little description nationally of core requirements; however, the Civil Contingencies Act 2004 clearly places a duty upon chief officers for their organizations to be adequately prepared to respond to emergencies. It therefore follows that when a plan identifies and delegates roles and responsibilities to nominated members of staff there is a requirement to ensure that these individuals are provided with training to best equip them to manage the emergency effectively, an implicit requirement that these staff are fully competent to carry out the roles assigned to them.

Competence can be developed through experiential learning but should also be guided by training, testing and exercising, ensuring also that these become regular features of the organization's training regime and are embedded in induction and regular sessions. For example, EPRR best practice recommends that incident managers be exposed to regular training to remain competent and confident.

The two main challenges in delivering quality training in EPRR within the health sector are the size and the diversity of the responsibilities and the complexities of health care organizations, from emergency medical services to primary care including general practitioners and community pharmacies, and from large acute hospitals to small mental health providers.

An additional training challenge is the low frequency with which emergencies arise within individual organizations, therefore offering few chances to employ or practise skills. Much more likely are internal disruptions or business continuity events such as power failures, internal pressures across the health system or IT failures (see Chapter 10, this volume); therefore, it is important that training interventions take this into account and are relevant to both types of incident. Business continuity management (BCM) training is largely in the form of awareness and understanding the relationships and interdependencies within organizations; it is difficult to imagine a more complex BCM environment than that of an acute hospital delivering a number of essential and critical services simultaneously.

Assessing skills, knowledge and attributes requires a training needs analysis (TNA) (Fig. 11.1), which can be applied to individuals or teams in specified roles. A TNA should reflect any national requirements for emergency

(a)

Organization	CBRN Training Requirement (example)			
	Awareness	Level 1	Level 2	Level 3
Acute	X	X	X	X
District	X	X	X	
Community	X	X		
All other	X			

(b)

	Question	Yes	No	Rationale
1	Do we meet necessary standards?			
2	Is an incident likely to occur?			
3	Does our resource meet the perceived need?			
4	Can the training be delivered?			
5	Who needs what training?	Calculate numbers:		

Fig. 11.1. Two example training needs analysis grids (CBRN, chemical, biological, radioactive, nuclear).

management (by taking into account the needs of the wider health care system) as well as incorporating local elements identified through risk assessment. This will allow organizations to most effectively target limited training resources.

From these simple results, the size of the training programme can be identified. This can then be followed by a simple, second analysis of the delivery of training requirements, to answer five key questions: by whom, where, when, what frequency and at what cost.

11.2.2 Training delivery

Training for emergencies is mostly focused on the response element, which in the majority of cases will be short lived. The longer response commitment is to business continuity incidents, which may last for a considerable period of time. It is often forgotten that a business continuity response should mirror that of an emergency response and be managed by an appropriate level of management. Given that the formula for managing both events is the same, a greater training value to the organization can be achieved by providing combined, joint training.

The model of incident management in the UK, integrated emergency management (IEM), has preparation and response as two of its core elements. *Preparation* incorporates defined and rationale training regimes. The *response* element focuses on the three-tier approach of strategic, tactical and operational (also referred to as 'gold', 'silver' and 'bronze'). It follows therefore that some training can be targeted at individual levels. Strategic and tactical levels are reasonably easy to formulate, revolving around incident management roles, responsibilities and defensible decision making.

In the operational environment, training interventions are more a matter of processes and procedures, and focus on escalation measures up-scaling resources. Non-clinical managers should focus on the organization of non-clinical staff to deal with support functions such as media enquiries, finance and supporting friends and families. This will ensure clinical staff can deliver the operational medical response. It is essential that clinical and administrative groups train and exercise together.

Within the UK there are a number of effective training tools for emergency planners and responders which cover a range of delivery methods. The first is direct teaching and includes training in the areas of Strategic Leadership in a Crisis (SLC), Surviving Public Enquires and media training. NHS England facilitates the provision of these to NHS organizations in England. There are also a number of undergraduate and postgraduate courses available in the UK to enable staff to gain professional qualifications in EPRR. A second teaching method is distance learning via online learning portals and includes more specialist training such as chemical incident management, biological incident management and decontamination.

Additional means of delivering training at low cost include focused 'one to ones' and mini scenarios that can be delivered by e-mail and requiring directed answers. For larger audiences, merged training sessions can be developed using mixed groups of staff across organizations or through combining similar topics. Joint aims and objectives can often be achieved, particularly across the health economy, so avoiding duplication of effort. The formation of a training subgroup as part of the Local Health Resilience Partnership (LHRP) or Local Resilience Forum (LRF), where it exists, is also of great benefit.

Good training outcomes are the product of many considerations; however, in the context of this chapter it is worthwhile highlighting two to support those required to deliver training who may not possess teaching qualifications. First it is necessary to know exactly what each element of the training is designed to achieve (i.e. the outcomes). This should then be followed by the aim (probably best defined as 'what matters'). Often this is written as a broad strategic statement and is usually quite vague, but does show the reason for the training. Finally, the trainer should define the training objectives: how and what the delegates will learn from the sessions. In putting these essential elements together, the trainer is advised to refer to and make use of Bloom's taxonomy (Fig. 11.2). This pyramid diagram categorizes the six levels in which learning is achieved and which are important to consider when designing training for different roles or ability levels.

The trainer must also understand and accept that a single session of training is unlikely to be fully effective and thus a well-constructed training regime should be developed (such as that shown in Fig. 11.3). This clearly demonstrates that delegates will need to consider what has been taught to them (reflection), think about how this may help their future performances (conceptualization), before bringing all elements together in testing or training scenarios. Trainers should consider that, at the higher management levels of professional performance, managers could be expected to take personal responsibility for some of these elements before coming back to confirm the learning through group experiences.

Training and Exercising for Emergencies 113

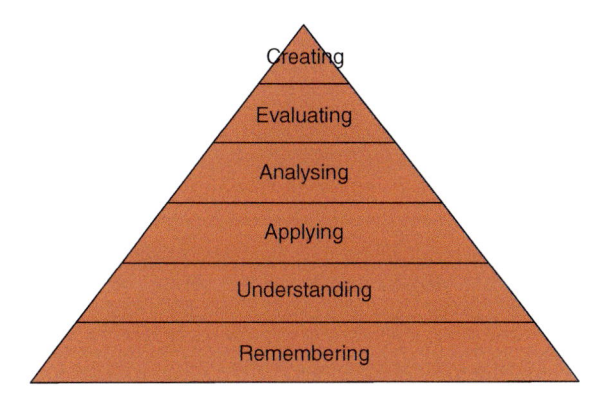

Fig. 11.2. Bloom's taxonomy (from www.learning-theories.com/blooms-taxonomy-bloom.html).

Fig. 11.3. Kolb's cycle of learning.

Figure 11.1. could be used to identify elements of specialist EPRR training in health care organizations. The development and use of an algorithm can be an excellent way of standardizing specialist training, but it is important to retain the human element for intervention in the form of past experiential learning. Training should empower staff members who work with other organizations to use their knowledge and skills to interpret individual problems and act accordingly, and in this way will complement algorithms with experience and good professional judgement. Their individual expertise should not be constricted by restrictive policies.

In the UK, it is important to remember that training is a specific requirement within the Civil Contingencies Act 2004. Additionally, any inquiry that may follow a major incident response is likely to ask two key questions of individual responders:

- What was *your* job during the incident?
- Were you *relevantly trained* to do it?

Scrutiny will always be applied; therefore, a final question that an inquiry will seek to address is:

- Could you have done a *better* job … if so how?

The training programme must clearly show how relevant training has been delivered if a response is required to these questions. A good starting point would be the National Occupational Standards developed under the Ministry of Justice programme, Skills for Justice. Twenty of these standards are directly applicable to civil contingency response, while a further 47 are 'relevant'. The NHS England EPRR Framework 2015 recommends these competencies for organizations coming under the guidance.

Those staff in organizations who are responsible for EPRR training must be inventive and realistic. Managers, particularly at executive/senior level, should assume some personal responsibility for their own learning.

11.3 Exercising

The use of exercises to test plans, processes, equipment and individuals' application of these is not a new development. The military were early adopters of using practical simulations to test aspects of their strategies and tactics. In emergency preparedness terms, an exercise is the practical simulation of a scenario which provides an opportunity to apply elements of a plan or emergency arrangements. Although an exercise will expose individuals to as close to a real situation as possible, this should not be seen as training. As specified earlier in this chapter, training should be provided to those responsible for mounting a response, to those following the plan, before they are required to implement it, whether for real or via an exercise.

11.3.1 Why is exercising important?

Exercising is another integral aspect of the emergency planning cycle. It helps us ensure that our arrangements are fit for purpose and safe to be implemented. A well-planned and well-run exercise will facilitate the testing and evaluation of plans, policies and procedures to help identify planning weaknesses, gaps in resources and capabilities, and clarify roles and responsibilities.

A plan can be in place for a number of years and therefore it is important to evaluate arrangements through exercises. This helps ensure plans remain accurate, up to date, workable and user-friendly. Organizations change, risks change, mitigating actions can change and resources available to manage emergencies change; therefore, it is important to not just update a plan on a regular basis but also to test it to ensure that it remains current and viable. Individuals within organizations change and a great deal of 'corporate memory' also leaves; testing and exercising finds these gaps. Equally, a plan can be understood by its author but needs to be tested by those responsible for implementing it to ensure that the actions are clear and easily followed.

Exercising new plans, policies and procedures helps reveal any gaps that the planning process has not identified and suggest possible alternative courses of action that had not previously been considered.

Ultimately an exercise can be used to validate and sign off a plan from an organization's perspective; to be able to declare that a plan has been tested by a lifelike scenario is a strong statement to the organization and the public whom the plan would serve.

It is important to remember that exercises should not be seen as an opportunity to test individuals but instead are intended to challenge plans, processes and arrangements. It is therefore important to consider the pressure that an exercise can place individuals under. A 4-day international exercise one author was once involved in had to be aborted a day early when the emotional impact on the players of testing the plan to breaking point was recognized.

11.3.2 Types of exercise

There are five recognized exercise formats that can be adopted depending on the plan or aspect of a plan that needs to be exercised. The five formats can be grouped into two distinct groups (Fig. 11.4). This is further elaborated in Table 11.1.

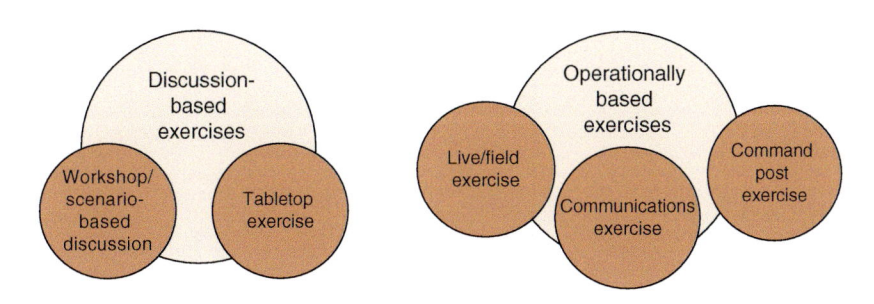

Fig. 11.4. Discussion-based and operationally based exercises.

Table 11.1. Descriptions of discussion-based and operationally based exercises.

Exercise type	Why?	Where?	How?
Workshop/scenario-based discussion	To familiarize staff with the organization's emergency preparedness and response arrangements, including plans, procedures, standard operating procedures, etc. To help develop a plan or process through a guided discussion with those likely to implement the plan	Usual workplace, meeting room or conference room	Using a plan or scenario to induce and guide thinking, discussing, problem solving, etc.
Tabletop exercise	To test and evaluate plans at the strategic and tactical levels in a safe environment, using a scenario or narrative as a trigger for discussion	Meeting or conference room	A structured scenario-based discussion with injects, triggers and background information to induce interacting, thinking and discussing as situational information is conveyed
Communications exercise	To test hardware as well as staff familiarity with processes, response times, inter-organizational cooperation, resources and contact information	Within the organization workplace, control rooms, employees' homes, on-call processes	A physical alert or cascade of information across a number of people, sites and/or organizations
Command post exercise	To test and evaluate the capabilities of the organizational emergency command and control response	Emergency operations centre or equivalent	Testing the organization's full command and control procedures and staff members' ability to move to their emergency roles and responsibilities from their routine activities
Live/field exercise	To test and evaluate a specific part of the emergency plan by mobilizing personnel and resources under live or field conditions To test and evaluate the use of equipment and physical resources	In a simulated environment with props and physical resources	Exercising a scenario in real time using simulated casualties, real-time information and real equipment

Choosing the right type of exercise will depend on a number of factors, including:

- *The purpose of the exercise.* What are the aim and objectives of the exercise and what is the desired output? Developing staff understanding of the business continuity plan may best be done via a workshop or discussion-based exercise. Testing the process to decontaminate casualties would best be achieved via a live exercise.
- *The target audience.* The target audience will also dictate the most appropriate exercise type: testing members of the executive team in their understanding of their roles and responsibilities in an emergency could be achieved via a tabletop exercise, while exercising control room staff may best be completed via a command post exercise.
- *The budget.* With a limited amount of money, a discussion-based or communications exercise can provide a test without needing an extensive budget. With more finance available, a comprehensive live field exercise with actors and additional staff overtime can be afforded.
- *Time available.* The time taken to plan for a good exercise should never be underestimated. A good live exercise can take up to 10 months to plan and deliver, where in contrast a communications exercise can be achieved with a few hours' preparation. Above all it is important to give participants as much notice as possible so as to guarantee their attendance.
- *Risks arising as a result of conducting the exercise.* Undertaking exercises can pose a physical and reputational risk to individuals and organizations. Live exercises could place individuals into hazardous situations such as collapsed buildings and crashed vehicles. Equally, a tabletop exercise attended by a number of external organizations (including the media) could have a reputational impact on an organization if the plan was not adequately embedded.
- *Available resources.* An exercise without the appropriate resources will not deliver its objectives. A severe weather tabletop exercise will fall short if there is inadequate weather forecasts and mapping available to the players.
- *Maturity of plans.* A mature and well-embedded plan would benefit from robust challenge from a large tabletop or live exercise, while a less mature plan may be more appropriately tested via a workshop or discussion-based exercise.
- *Experience of planners.* An experienced emergency planner will feel comfortable planning and delivering a large tabletop or live exercise, while less confident staff will sensibly want to start small with a communications exercise and build confidence and experience to deliver a live or command post exercise.

11.3.3 Exercise development

The first point in the exercise development process is to be clear about why we are undertaking the exercise and what we are aiming to achieve.

Towards this, it is important to set the aim from the onset. Examples of an exercise aim are:

- to evaluate or rehearse certain procedures;
- to test whether group responsibilities are known, understood and correspond with the plan;
- to test or rehearse the interaction between departments, organizations and agencies involved; and
- to validate plans and procedures.

To support the aim there should also be a number of objectives. These may differ for different departments or organizations; however, they should be overt and communicated to all planners and players early on in the process. It is important not to have too many objectives and to ensure they do not drift during the planning process. As planners get involved and are keen to make the best of the exercise opportunity, there can be a tendency to keep adding objectives. This will only serve to make the exercise more complicated and increase the possibility of not achieving the desired outcome.

Exercise objectives should be:

- set as targets for the exercise to achieve;
- consistent with the overall aim of the exercise; and
- follow the SMART principles of being specific, measurable, achievable, realistic and time-bound.

Specific and targeted aims and objectives will ensure a greater chance of the exercise achieving them. As the planner (or planning team) progresses through the planning process it is important to constantly refer back to them to ensure focus and prevent 'mission creep'.

11.3.4 Developing scenarios

Once the aim and objectives are set, the scenario can be developed. The best approach to developing a good and believable scenario is to base it on a real incident. This will ensure that exercise players believe the scenario and are therefore more likely to participate fully. A simple search on the Internet will provide many media news reports or even post-incident reports that can be used to develop a believable scenario. Contact with peers will also help as they may have exercised the same objectives and a previously developed scenario can be 'borrowed' and adopted.

As the scenario develops, a storyboard with key variables and parameters will start to form. This provides a narrative which sets the scene throughout the exercise and the exercise time frame. This may be set over a few hours, such as a live casualty management exercise, or a number of months in the case of a pandemic influenza tabletop exercise. To ensure that the exercise moves forward at the required pace and that key issues are tested or discussed, injects can be introduced. These can be delivered in a number of ways: ranging from a verbal briefing or an e-mail, to a weather report or something as elaborate as a simulated explosion or fire.

These provide the exercise controller with the ability to steer actions, slow or speed up play, or move people away from discussions or actions that are distracting and not meeting the exercise aim. The majority of injects should be pre-planned and timed so that exercise controllers know what to expect and when. That said, it is always worth having some hot or contingency injects in reserve so that exercises can be brought back on track if they start to drift or if play runs out of energy. In the exercise development process, it is worth spending time to consider where exercise play may slow or divert away from the objectives; this will provide the context for developing hot injects. Although it is acceptable to develop hot injects at the time, it is important to have a clear process and resource to do this or it may appear ad hoc and ill planned by the players.

Once the scenario injects have been developed, these all need to be held together by a master events list (MEL) (Fig. 11.5). This will provide the exercise controller and facilitators with a clear list of actions which the exercise requires to achieve its aim and objectives. For small communications exercises this can be as simple as a list of numbers to call and when, while for a full live exercise this can be a comprehensive list of all the agencies' actions, triggers and injects. This MEL is vital for all sizes of exercise; even when delivering a small exercise single handed it ensures that key actions, conversations and injects are not missed.

11.3.5 Exercise roles

It is a team effort to deliver a good and effective exercise. Clearly, the larger the exercise the more staff required to deliver it; the smaller the exercise the smaller the team. There are some key roles that are identified to deliver an exercise, as follows.

Exercise director
The exercise director is the individual with ultimate responsibility and accountability to oversee the whole exercise and all its functions. The exercise director has the overall responsibility and authority for starting, stopping or pausing the exercise, as well as managing the exercise controller, facilitators and observers (see below). He/she will also monitor progress against the MEL and can authorize the use of hot or contingency injects.

Exercise controller
The exercise controller is used predominantly in an operationally based exercise or combined with the exercise director for discussion-based exercises. The exercise controller provides exercise direction and control while following the sequence of events, and provides players with material according to the MEL. With the authority and direction of the exercise director, the exercise controller is able to intervene in exercise play to ensure that the objectives are met when play goes off course. In large exercises the exercise controller will be based in an exercise control room (EXCON) with a team of people to enact actions on his/her behalf.

Event number	Time	Title	To	From	Controller	Method	Type	Description/script	Expected Action	Comments/notes
1	09.00	SIMCELL operational	All controllers	SIMCELL	SIMCELL	Phone	Contextual	Simulators are in position and communications are in place	Check communications with field controllers	
2	09.45	Communications check	All controllers	SIMCELL	All	Radio	Contextual	Ensure communications are operable	All controllers establish communications with the SIMCELL	
3	10.00						STARTEX			
4	10.01	Report from train passenger	Salt Railroad Station security	Passenger 1	SIMCELL	Phone	Contextual	'This is Mary Banks at the Salt Railroad Station. I'm calling to report three suspicious people hanging around the train yard. Two of them are carrying backpacks. Please send someone immediately.'	Call Salt Springs Dispatch	

Fig. 11.5. An example of a comprehensive master events list.

Facilitator/umpire/directing staff
For discussion-based exercises there can be table or group facilitators embedded in with the exercise players. Umpires or directing staff work with the exercise participants to help to problem solve, keep participants on track, control the pace of injects, deliver messages and draw out key answers or solutions. These people are the key link between exercise play and the exercise controller. The umpire is able to spot periods of inactivity or loss of focus from the players and can therefore request the use of hot injects.

Evaluator
Exercise evaluators are there to capture the learning from an objective and informed perspective. These individuals are often selected on the basis of their expertise and impartiality. Evaluators should be provided with a framework or template to help identify and record the achievement of aims and objectives, as well as the actions of the players in response to injects. Ultimately the evidence provided by the evaluators will become a large part of the post-exercise report. There can be a real temptation to combine the umpire and evaluator roles into one; however, this should be avoided if possible. Whereas the focus for the umpire is to keep the exercise on track in order to deliver the aim and objectives, an impartial evaluation can only be provided by someone that has no other distractions or responsibilities.

Observers
Observers should be invited guests with no official role in the exercise and therefore they do not participate in the player activities or get in the way. For large or high-profile exercises there may a large amount of interest in watching and observing the exercise; the observers therefore require hosting by a dedicated member of staff and will follow a defined observer programme. Observers should be asked to provide feedback based on their observations and this may follow a pre-supplied format and support the production of a post-exercise report. Where logistics are difficult or the exercise is of a sensitive nature, it may not be possible to accept observers. In these cases, the production of a video may help share exercise learning in a more manageable manner.

Subject matter experts
Subject matter experts should be available to provide expert advice and guidance where the scenario requires. In the example of a tabletop exercise testing the response to a deliberate release of anthrax, it will be vitally important that there is an expert available to answer the technical questions that may arise. As technology progresses these individuals may not necessarily need to be present at the exercise. It may be possible for players to telephone, e-mail or video conference these experts so that they can provide the information remotely.

Exercise control

EXCON is a facility for large exercises to effectively drive the exercise during play. This facility can feed written and verbal information into exercise activity as well as monitor progress against the MEL. The larger the exercise the larger the EXCON is likely to be and will require the relevant technology such as e-mail, telephony and a video link to the exercise play or discussion.

Players

The term 'players' is used to describe those who take part in exercise activities. Players are present to act out the role they have been designated in the event of an emergency. Players are required to respond to information as they would in real life in order to simulate the actual response to an emergency. Without players, an exercise will not happen.

11.4 Delivering the Exercise

Ahead of exercise play a pre-exercise briefing should be provided to all participants, both delivery staff and players. The briefing should include information such as:

- exercise aim and objective;
- exercise scope;
- exercise format and timings (especially exercise start, stop and lunch);
- ground rules and code of conduct;
- individual and organizational roles and responsibilities (including exercise staff's parameters and limits of play – what is in and out of scope);
- simulation activities (what is real play and what will be simulated or imagined);
- exercise evaluation and debrief arrangements; and
- a brief overview or narrative giving a description of events occurring prior to the exercise start to set the mood and provide information the participants will need during the exercise.

Where possible this information should be provided ahead of the exercise. This does not always have to be delivered in person but can be provided in joining instructions, an e-mail, a teleconference, a video conference or a pre-prepared video clip. This and other relevant information can be supplied to staff and players in an exercise participants' handbook. This can include aspects such as staff and player procedures and responsibilities, safety and security requirements, communications methods available on the day, pre-exercise reading, training and evaluation process, etc.

Good exercise direction and facilitation will ensure that the exercise follows the agreed course and that the aim and objectives are met. It is sometimes necessary to pause or stop an exercise continuing. This may be as a result of the exercise moving in a direction that does not fulfil the aim, because the safety of participants is at risk or because a real incident has occurred. To support this and other activities, a set of common exercise code words is recognized (Box 11.1).

> **Box 11.1.** Exercise code words.
>
> - NAME: all exercises should have a code name to identify play from business-as-usual activities (e.g. Winter Willow, Common Ground).
> - STARTEX: start of exercise.
> - HOLD: suspend exercise for a period.
> - RESUME: start again after a hold.
> - SAFEGUARD: real incident/message outside of exercise play.
> - NODUFF: real emergency/casualty.
> - ABORT: early termination.
> - ENDEX: end of exercise.

In order to give an exercise an identity that planners and players can recognize, many are now given a code word. These are words often related to the subject of the exercise without giving any detail away. An NHS Olympic exercise called 'Exercise Milo' was named after an ancient Greek boxing athlete.

11.5 Exercise Evaluation Plan

The evaluation process should be designed at the same time as developing the exercise. To simply deliver an exercise is not enough; there must be a mechanism to capture the learning so as to embed into preparedness. After all, the purpose of an exercise is to test, evaluate or validate a plan and therefore an evaluation process will be needed to evidence that this objective has been met. The evaluation process can include a number of different aspects, such as:

- debriefing players and exercise staff;
- written responses to injects;
- reports from the evaluators;
- completion of player evaluation forms;
- observations from observers; and
- recorded activities (audio or video).

The evaluation process helps to ensure that a post-exercise report is prepared, agreed and circulated. The only way that learning will influence and change practice is if it is written up and shared. The post-exercise report should be based on fact, and be concise, objective, and prejudice- and blame-free. Learning can be identified without pointing the finger at an individual or organization. An example of a post-exercise report structure would include:

- an executive summary;
- an introduction;
- the exercise aim and objectives;
- an outline of the scenario;
- the exercise format;

- the exercise evaluation process;
- the lessons identified and suggested action; and
- a summary.

11.6 Summary

A well-prepared and well-managed emergency plan will contain the organization's training, testing and exercising regimes as an appendix within the administration section of the plan.

Training is an essential element of the EPRR cycle. The logical step after training staff to respond to a scenario is to use exercises to test plans, help identify gaps in capability, bring about new ways of resolving issues, expose teams to new scenarios and help validate planning. A well-run exercise will help evidence plans and put learning back into response.

Key Answers

- The response to large-scale emergencies is outside the normal experience of emergency responders. A paramedic will routinely respond to small road traffic collisions, but rarely to larger incidents where casualty numbers exceed the resources available. Equally, managers in strategic health care organizations will not be regularly exposed to emergency response management. Therefore, training to equip these people to effectively respond to and manage these situations is vital.
- Once the training need has been identified through a structured training needs analysis (TNA), training can be delivered through a variety of different methods. These include direct teaching, experiential or scenario-based leaning, and distance/online training. A blended approach to learning is often the most effective solution.
- The effectiveness of training can be assessed through a number of methods. This can be through formal assessment where evidence needs to be provided that the individual has retained the right amount of learning. Training effectiveness can also be assessed through practical demonstration such as exercise.
- Once a plan is written and staff are trained, it is important to evaluate the effectiveness of these arrangements. This can be tested under exercise conditions where a response can be evaluated against a particular scenario. There are five types of exercise format that can be used to evaluate preparedness: workshop/scenario-based discussion, tabletop exercises, live/field exercises, communications exercises and command post exercises. A range of features will dictate which type is most appropriate.
- Every exercise will identify lessons that can make a response more effective. These need to be presented in a post-exercise report so that others can identify the learning and incorporate it into their own emergency arrangements. Emergency response practice must be evidence-based and the vast majority of this comes from regular exercising.

Further Reading

Learning-Theories.com (n.d) Bloom's Taxonomy. Available at: www.learning-theories.com/blooms-taxonomy-bloom.html (accessed 1 March 2016).

Cabinet Office (2006) Emergency preparedness. Guidance on part 1 of the Civil Contingencies Act 2004, its associated regulations and non-statutory requirements. Chapter 5: Emergency Planning (revised October 2011). Available at: www.gov.uk/government/publications/emergency-preparedness (accessed 18 December 2015).

Federal Emergency Management Agency (2002) Summary of Post 9/11 Reports 'Lessons Learned'. Cross-Cutting Analysis of Post 9/11 Report 'Key Recommendations' for Improving the Nation's Preparedness. Available at: www.hsdl.org/?view&did=448479 (accessed 20 December 2015).

Kolb, D.A. (1984) *Experiential Learning: Experience as a Source of Learning and Development*. Prentice-Hall, Upper Saddle River, New Jersey.

NHS England (2015) *NHS England Emergency Preparedness, Resilience and Response Framework*. Available at: www.england.nhs.uk/ourwork/eprr/gf/#preparedness (accessed 30 December 2015).

NHS London (2009) *Review of five London hospital fires and their management, January 2008–February 2009*. Available at: www.preventionweb.net/files/13954_reviewoflondonhospitalfires1.pdf (accessed 20 December 2015).

Toft, B. and Reynolds, S. (2006) *Learning from Disasters: A Management Approach*. Perpetuity Press, Leicester, UK.

UK Commission for Employment and Skills (2011) *NOS Strategy 2010–2020, Revised Strategy June 2011*. Available at: https://www.gov.uk/government/uploads/system/uploads/attachment_data/file/304235/nos-strategy-2011.pdf (accessed 3 March 2016).

12 Post-incident Follow-up

K. Reddin[1] and G. Macdonald[2]

[1]*Strategic Emergency Planning Manager, Public Health England, London, UK*
[2]*Director of Studies, Organisational Resilience Programmes, Loughborough University, Loughborough, UK*

> **Key Questions**
> - What are the key aspects of learning lessons from actual incidents and exercises?
> - What are the benefits of learning lessons from actual incidents and exercises?

> *'Experience is simply the name we give our mistakes.'*
> (Oscar Fingal O'Flahertie Wills Wilde, 16 October 1854–30 November 1900)

12.1 Introduction

A key element of the integrated emergency management framework (which comprises the processes of anticipation, assessment, prevention, preparation, response and recovery) with regard to major incidents and emergencies is the implementation of a post-incident follow-up process or review in order to evaluate the response and recovery to an incident at any level of response.

'We will learn the lessons of this!'

This phrase is often used by politicians or executives in charge of departments or organizations that have been involved in the response to or management of a crisis, emergency or major incident where, more often than not, there is a considerable perception that all was not handled as well as it should have been. A corollary to this is that most UK health care organizations require, within their emergency preparedness, resilience and response (EPRR) arrangements, a lessons identified process to be implemented as part of their post-incident recovery procedures. This chapter considers some drivers and strategies that can be used to inform, supplement or further develop an effective lessons identified process.

An important question to address from the outset is 'why do we need a lessons identified strategy?' Arguably, it is a costly, resource-demanding and time-consuming process. In their findings, recent inquiries into major incidents have reported that learning identified in previous incidents and subsequent inquiries had neither been actioned nor was any effective change in behaviour or process implemented. So how do we ensure that lessons are indeed learned, resulting in a positive change in practices and behaviours, that will allow for efficient and successful responses?

To address the questions above, there are two fundamental areas (identified by Boin *et al.*, 2008) that are sensible to consider:

- *accountability* – mainly looking back and judging the performance of people; and
- *learning* – lesson drawing is more about looking forward and improving the performance of structures and arrangements (policies and practices).

Although learning is distinct from accountability, they may overlap in practice. Given the current level of post-incident investigation (official and media-led) into the management of responses and the prevailing perception of working in a blameworthy culture, it would seem natural to assume that those responsible for ensuring a responding organization has learned from past experience would be very keen to embed that learning into their planning processes before the next response.

So, if we accept that accountability and learning are two underpinning drivers for organizations to undertake a 'lessons process' post incident, how do we guarantee that, as identified by Coles (2014), we will not continually fail to convert 'lessons identified' from emergency responses into embedded 'lessons learned'?

Towards addressing this, a simple stepwise process can be applied to enable lessons identified to transition into recognized and embedded lessons learned (Fig. 12.1).

The rest of this chapter explores and reinforces the points made above to provide the reader with useful supporting information and guidance towards establishing an effective process for turning *lessons identified* into *lessons learned*.

12.2 Background

The scale and complexity of incidents, emergencies and natural disasters, nationally and internationally, have heightened the need to consider the management of knowledge and the importance of identifying lessons across different sectors, including health. In addition to incidents resulting from threats and hazards (including terrorism, natural disasters, accidents and industrial action), many organizations also have a commitment to take part in

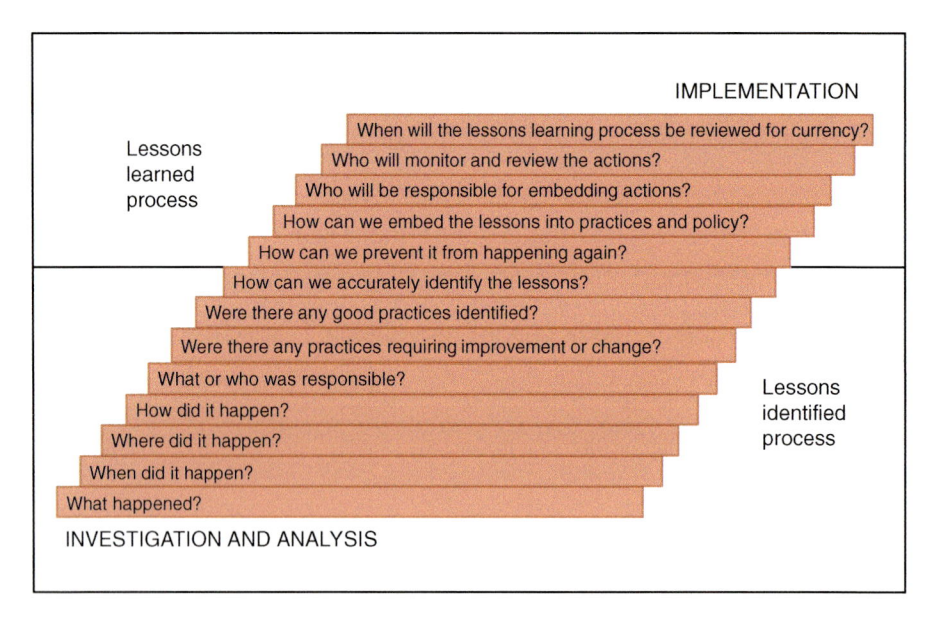

Fig. 12.1. The transition from lessons identified into lessons learned.

exercises to test their plans and arrangements for responding to such threats (Chapter 11, this volume).

Responses to real or simulated incidents will depend upon the nature and scale of the event and will be managed at appropriate local, regional or national levels. Local Resilience Forums (LRFs) in England bring together all of the key organizations which would respond in an emergency. This also provides a mechanism for the capture, collation and distribution of lessons or best practice in a multi-agency environment. In the UK, the Civil Contingencies Act 2004 requires all Category 1 responders to regularly exercise their plans and procedures in collaboration with other agencies, and to put procedures in place to record and capture lessons emerging from testing plans and procedures.

12.3 Lessons Management Systems

An effective lessons management system is important in providing organizations with assurance that learning is captured and implemented to improve performance. Any methodology employed to capture the lessons identified from responses and exercises must also put processes in place to ensure that these lessons are also learned. In order to achieve this, an organization needs to develop and implement both a lessons management system and a way of turning lessons into new policy and doctrine.

If these concepts are effectively employed the desired outcomes will be achieved, namely:

- embedded learning;
- positive changes in organizational processes;

- positive changes in behaviours; and
- development of a culture of organizational learning and change.

The benefits that may be realized by developing an effective lessons identified into lessons learned strategy, methodology and processes are:

- a collaborative approach that will encourage adaptability and flexibility across the organization and working with other sectors;
- continuous improvement of people within the organization;
- reduced operational risk, increased cost efficiency and improved operational effectiveness; and
- assurance to the relevant accountable officers that operational plans and processes are resilient.

Key to realizing these benefits and perhaps the biggest challenge to realizing the potential benefits is effectively engaging and sharing the lessons identified and how these have been addressed within and across organizations and jurisdictions.

12.4 Policy and Strategy Considerations

In the UK, a national policy and framework for identifying and learning lessons has been produced by the UK Cabinet Office which sets out a policy framework for capturing and implementing lessons at local, regional and national levels.

For the UK health sector and more specifically the NHS in England, the NHS England EPRR Core Standards set out, under the duty to maintain emergency plans and business continuity plans, that NHS organizations should have in place arrangements to carry out a debrief process so as to identify learning and inform future arrangements. There are also guidelines on the timescales for carrying out the debrief process.

12.5 A Process for Identifying Lessons

There are a number of methods that can be used to identify and capture the learning from emergencies and exercises. Most of these involve four key elements in the management of a lessons learned system:

1. Observations and data collection.
2. Analysis.
3. Implementation.
4. Monitoring and review.

Figure 12.2 provides a useful schematic of a lessons identified process that incorporates these elements.

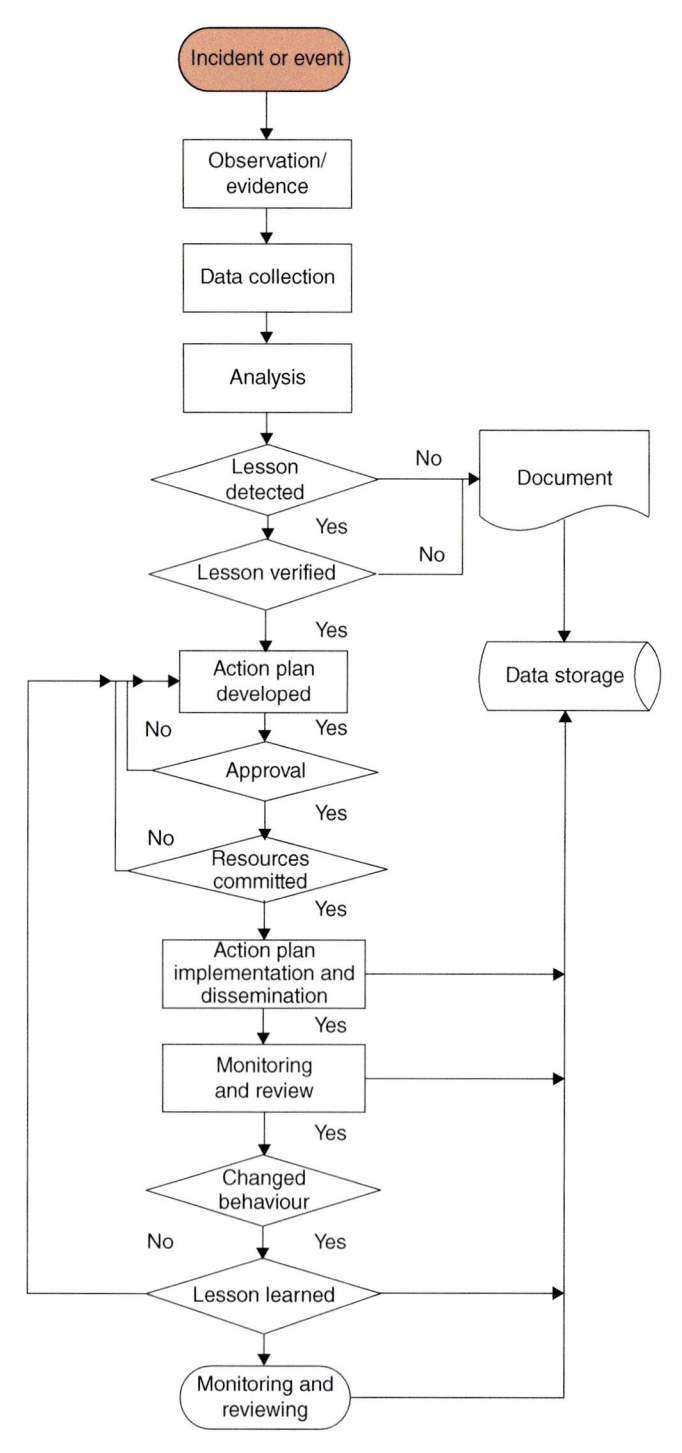

Fig. 12.2. Adapted schematic of a lessons identified process. (From Attorney-General's Department, Commonwealth of Australia, 2013.)

The process uses an integrated approach to provide a forum for those involved in an incident or exercise to express their observations and allow identification of:

- the principal issues;
- the root causes of these issues; and
- recommendations to address the issues and development of an action plan with clearly identified responsibilities and timescales.

The process of identifying lessons and capturing observations and other relevant data should begin as soon as possible during the response phase of the incident and carry on into the recovery phase. It is useful to identify time points during the response and recovery phases of an incident when a formal review of the lessons identified will take place; this will be especially important during a protracted incident. A time point should also be agreed at which to undertake a final formal review of the lessons identified based upon the observations and data captured throughout.

12.5.1 Observations and data collection

There are a number of recognized methodologies for capturing observations and data, with the main one being the post-event review or debrief. The types of debrief are as follows:

- *hot debrief* – this must be held immediately after the incident/exercise or once a shift or response is completed; and
- *cold debrief* – held some days, weeks or even months after the incident/exercise.

These can take the format of:

- *internal debrief* – involves staff within an organization only; or
- *multi-agency debrief* – involves staff from many organizations involved in an incident/exercise.

It is important to capture information as soon as possible after the event in a non-threatening, blame-free environment. If the incident continues to be managed over the medium to long term it may be necessary to hold regular debriefs at key milestones. Other methods of capturing data include virtual debriefs using surveys or questionnaires submitted by individuals and/or groups; and workshops and/or conferences convened to gather data. However, it is important to emphasize, whatever method is used, that it takes place at predefined time points during the response and recovery in the case of a protracted incident, or as soon as possible after the event or within a defined time frame, in order that:

- the people involved remember the events with some clarity;
- documentation relevant to the event can be retrieved easily; and
- it is likely that the people involved would still be with the organization, or traceable if they had changed their employer or retired.

Debriefs normally target specific themes that require investigation through structured debriefing; however, the lessons management process needs to be flexible enough to capture the unexpected. The themes that may be explored through the debrief process are often referred to as PPOSTT (people, process, organization, support, technology and training).

A facilitated structured debrief is used to bring together people involved in the response and should be structured to draw out learning, both positive and negative, encountered as part of the response to the incident or exercise. The PPOSTT themes can be further subdivided into categories in order to explore the issues that will need to be covered as part of the structured debrief. Some examples of categories are shown in Table 12.1; however, this list is not exhaustive and can be added to as required.

It is also important to have in place a stakeholder engagement and a collaboration strategy that treats the information exchange as mutually beneficial. The return for participating in an ongoing lessons management process is that all contributors see a tangible result for their input to the process.

The After Action Review process is a specific structured tool used by University College London Hospitals NHS Foundation Trust and others to identify and learn lessons, as well as to understand how or why they occurred (www.ucheducationcentre.org/behaviouralprogrammes.html). Originally developed by the US military, it asks three key questions:

- *What was expected?*
- *What actually happened?*
- *Why was there a difference?*

Before asking:

- *What has been learned?*

This is different from the more typical debrief process in that it starts by comparing what was intended with what actually happened.

12.5.2 Analysis

The key goal of the analysis of lessons captured is to determine the root cause of the observations and insights gathered from the debrief(s). From here, lessons can be identified and appropriate courses of action designed to embed the lessons can be recommended.

Table 12.1. Example debrief categories.

Coordination	Preparation	Communications	Resources
Internal	Internal	Internal	Staff
Multi-agency	Multi-agency	Multi-agency	Organization
		Media	
		Public	

The analysis step, which can be qualitative or quantitative, involves review of the collected data. However, the types of data collected through the debrief tend to lend themselves to qualitative methods of analysis as little of the data collected will be in quantifiable form. Martin and Turner (1986) argue that 'researchers must match research methods with research questions'. Grounded theory is particularly well suited to dealing with qualitative data of the kind gathered from semi-structured interviews and from case study material. The grounded theory approach offers the researcher a strategy for sifting and analysing material of this kind in order to identify trends or themes that an organization may use to define lessons to be learned. This informs possible solutions, strategy development and implementation options.

It is important to bear in mind that observations are people's perceptions. The analysis process needs to identify the factual evidence underlying these perceptions and identify the root causes of the situation. It is also important to remember that learning can be positive as well as negative. Root cause analysis, if applied correctly, will look at the evidence base with regard to what can be improved but also highlight areas of best practice that need to be shared. The most common root cause analysis tool used is the '5 whys technique' ('why, why, why, why and why') first published by Ohno in 1986.

12.5.3 Implementation

This stage considers the development of treatment options to recommend actions. Therefore, an action plan will be developed which should identify performance measures that will determine if the required changes have been implemented, and if they are effective and long-lasting. The action plan should set out how the implementation of any recommendations or actions will be verified, monitored and reviewed. Figure 12.3 provides an example of a useful lessons learned action plan template.

As shown, observations following an incident or exercise are not always negative; some may be positive. Therefore, the process should look at where practices need to be sustained as well as where they need to be changed, and verify, monitor and review all of these.

12.5.4 Monitoring and review

There are several ways to determine if a lessons management system is effective. Quantitative and qualitative measures can be used to assess things such as:

- changed behaviour or culture;
- increased operational effectiveness;
- better resource efficiency;
- improved safety;
- improved outcomes for the public and their communities; and
- increased compliance with policy, processes and procedures.

<<INSERT NAME OF INCIDENT>>

<<INSERT DATE OF INCIDENT>>

ACTION PLAN

Ref	Lesson identified/ recommendation	Means of delivery	Means of verification	Target timescale	Lead	Assumptions	Next steps	Change verified
001	Description of the lesson identified or recommendation	How this lesson/ recommendation is to be addressed and what actions need to be implemented	The evidence to show that actions are being addressed	Deadlines to include interim and final dates for implementation	Allocating responsibility to an individual, division or department who will be responsible for ensuring changes are implemented and reporting on progress, including completion	Additional information that might be important to complete the action, e.g. external factors	Information regarding work in progress or updates on the progress of actions	Provide details of how lessons identified/ recommendations have been verified and any evidence that demonstrates this, e.g. policy change, change in practice/behaviours

Fig. 12.3. Example lessons learned action plan template.

Monitoring the effectiveness of the implemented lessons needs to be carried out as outlined in the action and implementation plan. The results of this monitoring can indicate if implementation is on track or if changes need to be made to the plan.

Even when lessons have been verified as learned through evidence of changed behaviour, practices or procedures, this needs to be monitored and reviewed to ensure they are being sustained.

Reporting, monitoring and review helps develop a learning culture that is inclusive and sets the standards for implementing future lessons within the organization.

Reporting, monitoring and review should be an ongoing component of any lessons management system.

12.6 Summary

The key challenges to the health sector in responding to incidents or taking part in exercises is to ensure that the lessons identified become lessons learned. The key aspects and factors for success in learning from incidents and exercises are:

- *Sharing*. It is important to involve key partners and stakeholders in the learning process and to put in place an effective communications strategy. Gaining commitment from partners and stakeholders is vital to the success of the system.
- *Evaluation*. The process of producing a report to set out the key learning from each incident or exercise and subsequently reinforced by monitoring and review helps develop an inclusive learning culture. This process also sets the standards for implementing future lessons within the organization. Evaluation should be an ongoing component of any lessons management system.
- *Stakeholder agreement*. Agreement among all of the involved stakeholders of the lessons learned and action plan prior to implementation and embedding the learning within and across organizations.
- *Implementation*. Any lessons management system should also include a mechanism for producing an action plan out of the review or debrief which sets out how the recommendations and actions are to be implemented. However, the organization should be prepared to adapt this action plan if necessary in order to ensure that lessons are learned.
- *Embedding*. Within an organization, executive support and oversight is critical for successful implementation of any lessons management system and to ensure that learning becomes embedded within the organization. However, acceptance of the change in order to embed the learning needs to happen across the organization.

There are clear benefits in having a system in place that allows effective learning from involvement in incidents and exercises, particularly the development of a collaborative approach which will encourage adaptability and flexibility across the organization and working with other sectors. It will assist in the continuous improvement of people within the organization, which will in turn reduce operational risk, increased cost efficiency and improve operational effectiveness. An effective system for post-incident follow-up will also provide assurance to the relevant accountable officers within the organization that operational plans and processes are organizationally resilient.

The processes described in this chapter can be applied reasonably easily to help realize the benefits described; however, the key challenge will be to ensure and to demonstrate that the lessons identified have indeed been learned.

Key Answers

- Learning lessons is a distinct two-part process; initially we need to identify lessons from an incident or event and then convert them into lessons learned. Real incidents and exercises allow for analysis of actual experience and evidence of actions or processes applied to that event, be they good or bad. Analysis of the incident aids the identification of those lessons which, if applied on either an individual or an organizational basis, will improve or further develop areas of poor performance. Once lessons identified are implemented or applied to personal or organizational actions or processes and result in a positive change to practices, which can be evaluated through further exercises or indeed actual incidents, then and only then can they be considered lessons learned.
- Learning lessons from actual incidents and exercises encourages the development of a collaborative approach which should promote adaptability, flexibility, and continuous organizational and individual improvement. Additional benefits will lead to the reduction of operational risk, increased cost efficiency and improved overall operational effectiveness.

References

Boin, A., McConnell, A. and 't Hart, P. (2008) *Governing after Crisis: The Politics of Investigation, Accountability and Learning*. Cambridge University Press, Cambridge, UK.

Coles, E. (2014) *Learning the Lessons from Major Incidents: A Short Review of the Literature*. Emergency Planning College Occasional Papers New Series, Number 10. Emergency Planning College, York, UK.

Commonwealth of Australia (2013) *Lessons Management*. Australian Emergency Management Handbook Series, Handbook 8, 2nd edn. Australian Emergency Management Institute, Attorney-General's Department, Canberra.

Martin, P.Y. and Turner, B.A. (1986) Grounded theory and organizational research. *Journal of Applied Behavioural Science* 22, 141–157.

Ohno, T. (1988) *Toyota Production System: Beyond Large-Scale Production (English translation, ed.)*. Productivity Press, Portland, Oregon, pp. 75–76.

Further Reading

Cabinet Office (2008) Lessons Identified from UK Exercises and Operations – A Policy Framework. Available at: www.gov.uk/government/uploads/system/uploads/attachment_data/file/61349/lessons-learned-exercises-framework.pdf (accessed 5 August 2015).

Cabinet Office (2013) *Responding to Emergencies*. The UK Central Government Response. Concept of Operations. Available at: www.gov.uk/government/uploads/system/uploads/attachment_data/file/192425/CONOPs_incl_revised_chapter_24_Apr-13.pdf (accessed 5 August 2015).

HM Government (2004) *Civil Contingencies Act* 2004. Available at: www.legislation.gov.uk/ukpga/2004/36/contents (accessed 5 August 2015).

NHS England (2015) NHS England Core Standards for Emergency Preparedness, *Resilience and Response, Version 3.0, updated* 15 May 2014. Available at: www.england.nhs.uk/wp-content/uploads/2015/06/nhse-core-standards-150506.pdf (accessed 5 August 2015).

NHS England (2015) NHS England Emergency Preparedness, *Resilience and Response Framework, Version 2.0, updated* 10 November 2015. Available at: www.england.nhs.uk/wp-content/uploads/2015/11/eprr-framework.pdf (accessed 2 March 2016).

Toft, B. and Reynolds, S. (2006) *Learning from Disasters: A Management Approach*, 3rd edn. Perpetuity Press, Leicester, UK.

13 Mass Casualty Incidents

M. Shanahan

Head of Special Operations, Yorkshire Ambulance Service NHS Trust, UK

Key Questions

- What is a mass casualty incident?
- Why is it different from a major incident?
- What are the challenges for the NHS in managing this type of incident?

13.1 Introduction

This chapter largely considers the preparedness for and response to mass casualty incidents (MCIs) in the UK; however, principles can be extrapolated to other settings. The emergency services are well versed in planning and preparing for large-scale incidents. It is commonplace for events hosting in excess of 100,000 people to be planned for and safely managed. In the UK this is achieved through coordinated planning and management of the events using guidance such as the *Purple Guide to Health, Safety and Welfare at Music and Other Events* or *The Guide to Safety at Sports Grounds* (see Further Reading).

Apart from the planned events, the day-to-day business of the emergency services is to respond to the unplanned, no-notice incidents. In 2015 the police and ambulance services across England responded to over 22,000 emergency calls in any 24-hour period, with the fire and rescue services responding to approximately 1700 calls in the same period. A number of the calls related to multiple casualty incidents, arising from events such as road traffic collisions or relocating people who became displaced from their homes because of accessibility issues created by adverse weather events (i.e. flooding).

Conversely, MCIs generate significantly more casualties than the normal day-to-day business. The Department of Health (DH) defines the escalating numbers of casualties as:

- *major* – tens;
- *mass* – hundreds; and
- *catastrophic* – thousands.

The UK NHS definition of a mass casualty event is:

a disastrous single or simultaneous event(s) or other circumstances where the normal major incident response of several NHS organisations must be augmented by extraordinary measures in order to maintain an effective, suitable and sustainable response.

In the context of this chapter, MCI mainly refers to those patients who have sustained trauma-related injuries through either natural or man-made events. However, it should also be remembered that MCIs can be generated by events such as environmental extremes of cold, heat or flooding. The displacement of hundreds of people from their homes, while not an MCI in terms of injuries, is in terms of the scale of people involved. Many of them are likely to have left behind prescribed medication, or be in need of medical care for chronic illness, which would necessitate special measures being put in place.

NHS England states that:

The NHS service-wide objective for emergency preparedness, resilience and response is: To ensure that the NHS is capable of responding to significant incidents or emergencies of any scale in a way that delivers optimum care and assistance to the victims, that minimises the consequential disruption to health care services and that brings about a speedy return to normal levels of functioning; it will do this by enhancing its capability to work as part of a multi-agency response across organisational boundaries.

In the 21st century, society requires that all the services who would be involved in dealing with an MCI can respond timely, effectively and save the lives of as many people as possible. To do this requires a considerably different response from that which is delivered on a daily basis.

In 2011, the World Health Organization (WHO) stated that in recent years 1.2 million people have been killed in mass casualty events. In the 10 years prior to the report being produced more than 2.6 billion people had become casualties of natural disasters. Events such as earthquakes, landslides, cyclones and rising-tide events (e.g. heatwaves, floods or severe cold weather) (see Case Study 13.1 and Case Study 13.2) resulted in high numbers of casualties (Table 13.1). WHO projections suggest that these figures will increase by 65% over the next 20 years.

While in some cases we can predict natural disasters and so minimize their impact on human suffering and the cost to the economy in recovering from such devastation, we also live in a world that is bestowed with deliberate acts of violence, which create the same impact. Table 13.2 identifies some key deliberate acts of violence where at least 100 people have been injured since 2001.

> **Case Study 13.1.** Haiti earthquake, 2010
>
> 'The Haiti earthquake created 300,000 non-fatal casualties. Typically, approximately 60% of persons presenting to field hospitals require surgical intervention, of which 80% involved debridement of wounds and dressings with very few primary closures or external fixation procedures.'
>
> (From WHO Disaster Risk Management for Health Fact Sheet, May 2011)

> **Case Study 13.2.** Nepal earthquake, 2015
>
> On 25 April 2015 an earthquake in Nepal killed more than 8800 people and injured more than 23,000. It was the worst natural disaster to strike Nepal since the 1934 Nepal–Bihar earthquake. The earthquake triggered an avalanche on Mount Everest, killing at least 19 and another one in the Langtang valley, where 250 people were reported missing.
>
> Thousands of people were made homeless with whole villages being flattened across many areas of the country. Aftershocks continued throughout Nepal in 15–20 min intervals, with one shock reaching a magnitude of 6.7.
>
> On 12 May a major aftershock occurred, the epicentre of which was near to the Chinese border between the capital Kathmandu and Mount Everest. Two hundred people were killed and in excess of 2500 people were injured.
>
>> 'The repeated earthquakes and aftershocks since 25 April 2015 have had major public health consequences, with a total 1085 health facilities (402 completely and 683 partially) damaged. A total of 2088 people have undergone major surgeries and 26,160 have received psychosocial support in the highly affected 14 districts. 42 Foreign Medical teams (FMTs) are operating in the country with a total 802 persons including 264 doctors and 236 nurses. Currently there are over 100 beds available for patients requiring ongoing rehabilitation or nursing care within the Kathmandu valley.'
>>
>> (From WHO Nepal Situation Report # 19, May 2015)

Table 13.1. The largest natural disasters in recent years. (From *World Disasters Report 2010*.)

Date	Location	Event	Number of deaths	Number affected
Jan 2001	Gujurat	Earthquake	20,005	6,321,812
Aug 2002	Dresden	Floods	27	330,108
Summer 2003	Europe	Extreme heat	72,210	NR
Dec 2003	Bam	Earthquake	26,796	267,628
Dec 2004	South Asia	Tsunami	226,408	2,321,700
Jul 2005	Mumbai	Floods	1,200	20,000,055
Aug 2005	New Orleans	Hurricane	1,833	500,000
Oct 2005	Kashmir	Earthquake	73,338	5,128,000
May 2006	Java	Earthquake	5,778	3,177,923
May 2008	Indian Ocean	Tropical cyclone	138,366	2,420,000
May 2008	Sichuan	Earthquake	87,476	45,976,596
Mar 2010	Haiti	Earthquake	222,570	3,400,000
Mar 2011	Japan	Earthquake and tsunami	15,000	NR
Apr 2015	Nepal	Earthquake	NR	103,686

NR, not reported.

Table 13.2. Terrorist attacks since 2000 (only incidents with 100+ injured are reported).

Date	Location	Event	Number killed	Number injured
May 2001	Israel	Suicide bombing	5	100+
Jun 2001	Israel	Suicide bombing	21	100+
Aug 2001	Israel	Suicide bombing	15	130
Sep 2001	New York	Four planes hijacked	3000	Unknown
Mar 2002	Israel	Seven sites on different days	73	358
Jul 2002	Israel	Bombing	9	100
Oct 2002	Kuta, Bali	Bombing	202	240
Oct 2002	Moscow	Siege	170+	700+
Jan 2003	Tel Aviv	Shootings	23	100+
May 2003	Saudi Arabia	Shootings and car bombing	27	160
Nov 2003	Istanbul	Bombings	57	700
Mar 2004	Madrid	Train bombing	191	1,800
Jul 2005	London	Bombings (transport)	53	700
Aug 2007	Iraq	Bombings	796	1562
May 2010	Iraq	Bombings	100+	350+
May 2011	Egypt	Shootings	15	232
Jun 2012	Iraq	Multiple shootings/bombings	93	300
Jan 2013	Pakistan	Bombings	130	270
Apr 2013	Boston	Bombings	3	183
Sep 2013	Kenya	Shootings	67	175
Mar 2014	China	Knife attacks	28	143
Nov 2014	Nigeria	Bombing	120	260
Apr 2014	Afghanistan	Suicide bomb	33	100+
Nov 2015	Paris	Marauding terrorist firearms attack	130	368
Total			5261+	9131+

13.2 Pre-hospital Response

At any major incident and especially an MCI, the first hour of response is challenging. Information is usually received that is confusing and unclear and often contradicts information already received. It takes time to establish command structures, both at the scene and remotely, to ensure the information being received and passed is accurate. The ambulance service is the gateway to the NHS and it is they who coordinate the response at the scene and cascade the information from the emergency operations centre (EOC) to the receiving hospitals, using a predetermined message indicating the seriousness and the size of the incident and whether a major incident has been declared. The Ambulance Incident Commander (AIC) will be supported by a Medical Incident Advisor (MIA) who is usually a senior doctor appointed by the ambulance service and is responsible for the clinical care provided at the scene with the AIC. The AIC has overall responsibility for the medical response at the incident site, including the safety of all the medical staff. The AIC will work collaboratively with the police and fire and rescue service incident commanders.

The principles of managing a road traffic collision, a major incident or an MCI are the same. Key principles to extract those injured and convey to hospital are considered and implemented, using a structured approach to manage the incident. Since 2013 in the UK the Joint Emergency Services Interoperability Programme (JESIP) has defined the management approach that all the emergency services must adopt. This includes the National Decision Model (a structured approach to making informed decisions in a dynamic situation) and the Joint Dynamic Risk Assessment (joint assessment of the hazards and risks at an incident).

In addition to the above, another system that is used by the medical first responders is the Major Incident Medical Management and Support (Advanced Life Support Group, 2011). This system provides a sequential approach to scene assessment and management and the care of those injured. It is outlined in Fig. 13.1.

Medical resources that are sent to the scene in the UK usually consist of the ambulance service paramedic and emergency medical technician (EMT) personnel. As part of the command team there will also be an MIA, who will work closely with the AIC to advise on casualty numbers, severity of injuries and the most appropriate receiving hospital, based on their injuries.

A casualty clearing station (CCS) will be established near to the scene of the incident to manage the casualties coming from the scene, before going to a receiving hospital. There will be a combined team of ambulance staff, as well as wider health care staff, trained in pre-hospital deployments (Mobile Emergency Response Immediate Team (MERIT)), in addition to logistical support, to manage the flow of patients and replacement of equipment. The initial medical equipment needed to set up the CCS would be provided by the ambulance service. Where additional equipment is required, a mutual aid request to other ambulance services would be provided for the mass casualty vehicles held in each English ambulance service.

Most hospital emergency departments in the UK, even the large ones, can only manage between four and eight Priority 1 (P1) casualties at any one time. These are patients who have life-threatening injuries and require extensive medical, surgical and intensive care. In addition, the specialist beds such as intensive care, burns, paediatric and neurological beds are in small numbers comparatively and in different hospitals, invariably not geographically close by. With this in mind, the patients at the scene are likely to be held longer (particularly the Priority 2 (P2) patients – serious, but not life-threatening), stabilized, and moved to specialist centres or other hospitals outside the region to accommodate the P1 patients in the hospitals closer to the incident site. To do this safely requires additional skills and equipment not normally catered for by the ambulance service or the pre-hospital care schemes, such as the British

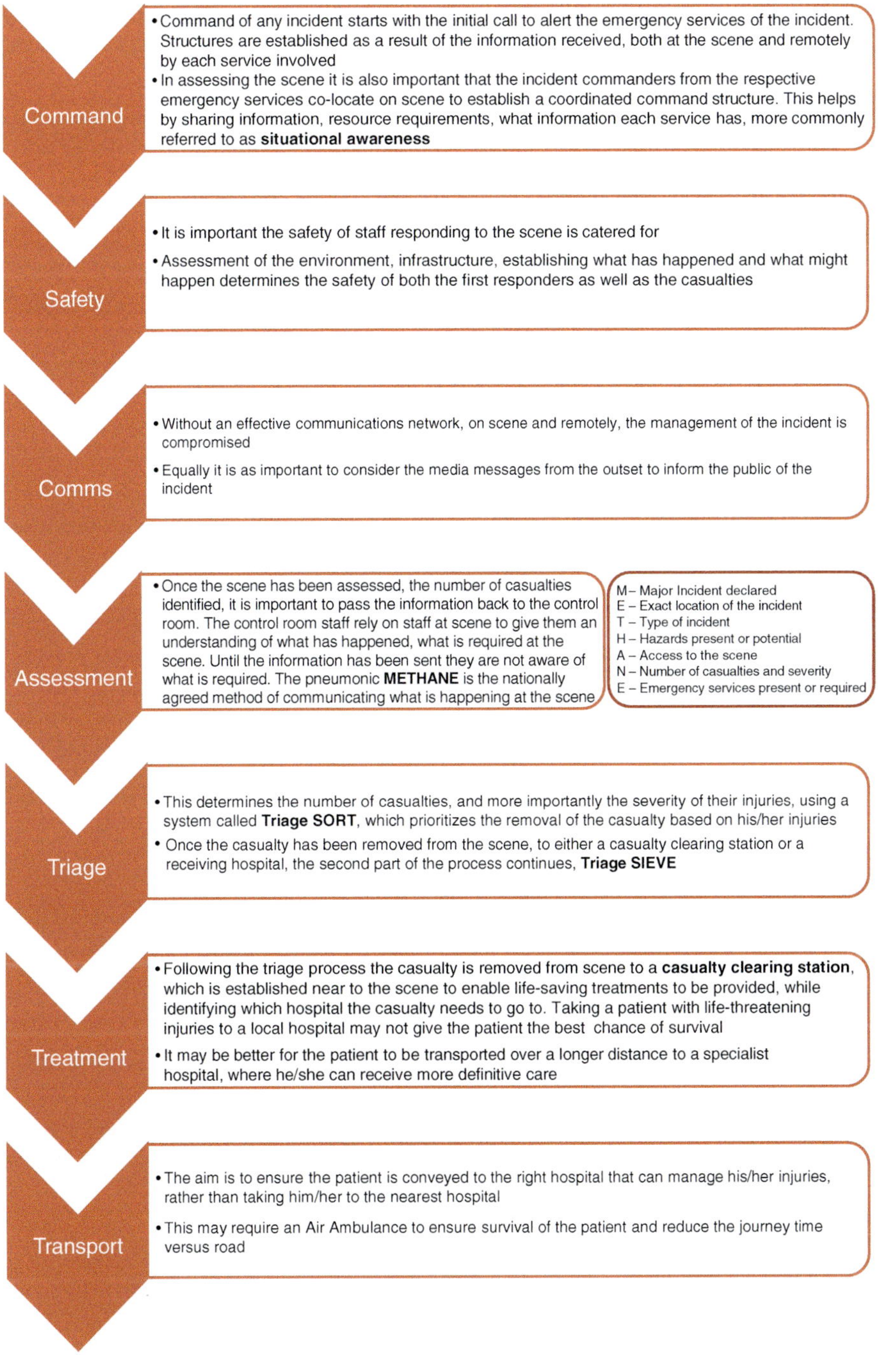

Mass Casualty Incidents

Command
- Command of any incident starts with the initial call to alert the emergency services of the incident. Structures are established as a result of the information received, both at the scene and remotely by each service involved
- In assessing the scene it is also important that the incident commanders from the respective emergency services co-locate on scene to establish a coordinated command structure. This helps by sharing information, resource requirements, what information each service has, more commonly referred to as **situational awareness**

Safety
- It is important the safety of staff responding to the scene is catered for
- Assessment of the environment, infrastructure, establishing what has happened and what might happen determines the safety of both the first responders as well as the casualties

Comms
- Without an effective communications network, on scene and remotely, the management of the incident is compromised
- Equally it is as important to consider the media messages from the outset to inform the public of the incident

Assessment
- Once the scene has been assessed, the number of casualties identified, it is important to pass the information back to the control room. The control room staff rely on staff at scene to give them an understanding of what has happened, what is required at the scene. Until the information has been sent they are not aware of what is required. The pneumonic **METHANE** is the nationally agreed method of communicating what is happening at the scene

  M – Major Incident declared
  E – Exact location of the incident
  T – Type of incident
  H – Hazards present or potential
  A – Access to the scene
  N – Number of casualties and severity
  E – Emergency services present or required

Triage
- This determines the number of casualties, and more importantly the severity of their injuries, using a system called **Triage SORT**, which prioritizes the removal of the casualty based on his/her injuries
- Once the casualty has been removed from the scene, to either a casualty clearing station or a receiving hospital, the second part of the process continues, **Triage SIEVE**

Treatment
- Following the triage process the casualty is removed from scene to a **casualty clearing station**, which is established near to the scene to enable life-saving treatments to be provided, while identifying which hospital the casualty needs to go to. Taking a patient with life-threatening injuries to a local hospital may not give the patient the best chance of survival
- It may be better for the patient to be transported over a longer distance to a specialist hospital, where he/she can receive more definitive care

Transport
- The aim is to ensure the patient is conveyed to the right hospital that can manage his/her injuries, rather than taking him/her to the nearest hospital
- This may require an Air Ambulance to ensure survival of the patient and reduce the journey time versus road

Fig. 13.1. Major incident medical management and support.

Association of Immediate Care Schemes. Consequently, the MERIT staff are brought to the scene with the advanced clinical skills and equipment. In some countries this can be referred to as a field hospital (see Case Study 13.3).

The welfare of all the staff working at the scene, or in the CCS, must also be considered. The staff sent to the incident site initially will need to be stood down for a period of rest, so a key consideration in the first few hours of any incident is the relief staff to take over from the first responders. This is likely to go on for many hours, so it is highly likely there will need to be additional staff brought in (at all levels of the response) for greater than 24 hours. Sufficient rest periods, feeding stations, sleeping arrangements, as well as the logistics of the equipment needed must all be considered and put in place. This whole support and logistics mechanism cannot be undertaken by the command teams trying to resolve the incident. It must be managed by a separate team and it should not be underestimated how complex and protracted it will become.

Specialist teams are likely to be sent to the scene to provide support if their skills are relevant to the incident. The UK Ambulance Service has the Hazardous Area Response Team (HART) and the Fire and Rescue Service has the Technical Rescue Teams, both capable of providing care and extrication at height, underground, in confined spaces, in non-oxygen atmospheres and in water.

Non-governmental organizations such as St John Ambulance and the Salvation Army, more commonly known as the Voluntary Aid Societies (VAS), will provide valuable assistance with all aspects of an MCI including trained medical staff, vehicles, individual registration and tracking, temporary shelter, food service, and many other important services. For organizations such as the VAS, who are predominantly volunteer staff, the logistics

---

Case Study 13.3. London bombings, 7 July 2005

On 7 July 2005, terrorists exploded a series of bombs in central London. The London Underground system was their target, exploding bombs at Aldgate, Kings Cross and Edgware Road underground stations. At approximately 09.47 hours the fourth bomb exploded on a bus in Tavistock Square. There were 54 passengers killed and 700 injured.

'54 are dead out of an estimated 700 casualties (7.7%). Approximately 350 were treated on scene and 350 transported to hospital.'

'Our helicopter was essential to deploy staff and equipment (but not patients) to the various scenes. It was also used to deploy staff to the hospital from home. It flew 21 sectors in the morning and allowed rapid deployment in gridlocked traffic conditions.'

'Scene safety is a major concern for rescuers. Any of these scenes may have contained secondary explosive devices. This is a well-recognised risk at terrorist incidents.
In addition, risk of structural collapse, inhalation of airborne particulate matter and contamination were also issues of concern.'

'The emergency services have to deploy extraordinary resources to major incidents, deal with the personal, administrative and media burden in the hours and days that follow and provide a comprehensive emergency service the next day. This requires careful resource management.'

(Extracts from Lockey et al., 2005)

of providing this level of support takes time to mobilize and invariably will come from many different areas. It is therefore important that the request for VAS staff and equipment is made early to give them as much time as possible to respond effectively.

The media plays an important role in informing the general public about the incident and in asking them to stay away from the incident site, but also in advising which roads are inaccessible or are being maintained for the emergency services.

Inevitably there will be casualties who do not survive or are found dead *in situ*. While the living take priority, it is important that the dignity of the dead is catered for. A temporary body-holding area should be established for the deceased. This also aids the coroner's officers and pathologists in their evidence-gathering procedures and establishing the cause of death. Ultimately, friends and family of the deceased will want to know what happened and what was done for their loved one after the event.

Given the number of casualties, it is highly likely they will be conveyed over some distances, especially if specialist beds are required as previously mentioned. In some cases, conveyance to a European Union country for the specialist care may be required. This would be coordinated through the regional offices for NHS England and the DH.

13.3 Survivor Reception Centres

Survivor reception centres (SRCs) are established to manage those people affected by the incident but not injured, or who are displaced from their homes and need shelter. Invariably these are schools, public buildings, sports or community centres, and are listed with the local authorities as being able to provide such a facility. For example, in the case of the Paddington rail disaster in 1999, a local supermarket was used to initially care for survivors.

The SRC is set up and managed by the local authority, with support from other agencies, such as the police, health service and the VAS. The SRC provides food and shelter for those affected by the incident, medical care where needed and support from the police in being able to identify relatives who may be missing. It is important these are activated early, as it takes time to establish the infrastructure needed. Activation is through the local authority (usually requested by the police) and would normally come through the multi-agency Strategic Coordinating Group (SCG) or the Tactical Coordinating Group (TCG) (see later).

13.4 Incident Coordination

The key to a successful outcome of any type of incident is the effective planning and preparation that occurs before the event. In the UK the Civil Contingencies Act (2004) and its underpinning guidance (the Emergency Response and Recovery Guidance, updated in 2013) aims to establish good practice, based on lessons identified from previous events, both in the UK and internationally. The guidance is designed to provide a framework for all agencies involved in the response and recovery phases of an incident. There are two types of responder, Category 1 and Category 2.

- *Category 1 responders* are those agencies immediately involved in the response phase to an incident: police services; fire and rescue authorities; ambulance services; Acute and Foundation Trust Hospitals; NHS England; Public Health England; Port Health Authorities; the Maritime and Coastguard Agency; local authorities; and the Environment Agency.
- *Category 2 responders* are those agencies involved in the support to the Category 1 agencies: utilities; telecommunications; transport providers; the Highways Agency; the Health and Safety Executive; Category 2 responder health bodies; and the wider resilience community.

During an incident a multi-agency SCG is established and in some cases a TCG. The TCG is tasked with managing and resolving the incident itself, whereas the SCG ensures the TCG has the resources to manage the incident and ensures its actions are compatible with the strategic aims set by the SCG to resolve the incident. In addition, the SCG considers the wider implications of the incident and its effect on the communities. The SCG feeds into Government departments and ultimately into the Government's Cabinet Office Briefing Room (COBR) (see later).

Depending on the incident type, a Scientific and Technical Advisory Cell (STAC) may be established to provide advice and guidance to the SCG on the impact on public health. A STAC is set up in the event of an emergency where there is likely to be a requirement for coordinated scientific or technical advice. Its role is to advise the SCG (and COBR) of the most effective way of dealing with contamination or hazardous materials, the impact to the environment and communities. For example, chemical contamination of an area would necessitate the STAC being set up. The STAC is usually chaired by Public Health England.

The VAS provide invaluable support during major incidents. They bring support to the statutory authorities, by providing some of the following functions:

- support to local authorities in providing emergency shelters, administrative support and psychosocial support;
- first aid at temporary treatment and emergency shelters;
- food for the public, emergency workers, volunteers and survivors at the SRC;
- administrative support for varied functions, including contact tracing for relatives;
- assistance with transport of patients and survivors;
- telephone helpline information;
- assistance following hospital discharge;
- access to mobility equipment for patients (wheelchairs, walking sticks, etc.); and
- assistance with providing longer-term care and support for recovery.

13.5 The Acute Hospital and Wider Health Care Response

When it is determined that the scale of the incident is such that it requires special measures to be implemented (i.e. declaration of a major incident), the hospitals in the immediate vicinity will be notified by the ambulance control room. They are informed via a standard message format, in the same way across all the hospitals. Based on this information the hospitals are likely to declare a major incident and invoke their major incident plan, or in this case the MCI plan. The difference between the major incident and the MCI is the ability to manage the numbers of casualties that will be arriving, which is achieved by actions such as suspending elective surgery lists, discharging patients early (where feasible) to free up beds, transferring some patients to other hospitals, calling off-duty staff into work, suspending outpatient clinics to provide capacity for minor injuries, making space for the relatives and managing the media interest.

In England, at the time the hospitals are notified, the wider health care organizations are also activated. NHS England is informed, which informs the DH. This ensures the whole system is aware of the incident. In large-scale incidents, particularly where mass casualties are involved, the government emergency committee (COBR) would be established. DH representatives form part of the committee and as such would be seeking information of the events taking place, particularly casualty numbers and receiving hospitals, so they can brief the Government representatives accordingly.

Because of the number of casualties being brought into the hospitals (see Case Study 13.4), those in the immediate area of the incident (which could extend across the local health geography) would start to clear their emergency departments and outpatient clinics and look to discharge non-critical patients to other hospitals or places of care, to allow for the influx of casualties. Discharges would go to a community bed (transition or step-down facility), a residential or nursing home, discharge to home with a domiciliary care package, or transfer to a temporary place of safety. Early engagement with the local authorities to identify nursing and residential beds would be established through the hospital command team.

Elective procedures would be cancelled to free up operating theatres and surgical beds. Additionally, diverts to other hospitals for patients coming into hospital through the business-as-usual emergency ambulance service activity may be secured to reduce demand on the receiving hospital.

The ambulance service would send a Hospital Ambulance Liaison Officer into each of the receiving hospitals to work closely with the bed manager, thereby managing capacity to demand and by specialism. This information is then fed back to the MIA on scene and the ambulance control centre. The emergency departments would establish teams to receive the patients at the door and commence their triage process, in the same way that it is done at the scene.

Hospitals outside the immediate area of the incident may be asked to provide medical staff to attend the scene and support the ambulance service (MERIT). Some areas have pre-hospital trained doctors who volunteer their

> **Case Study 13.4.** Mumbai terrorist attack, November 2008
>
> On 26 November 2008, terrorists entered Mumbai via the Cuffe Parade fishermen's colony, divided into five pairs and went to different locations to carry out their atrocities. The first pair went to the Chhatrapati Shivaji Terminus (CST) railway station. They entered the railway station and indiscriminately fired upon the people therein, placed explosives around the terminus and attacked with hand grenades. The second pair went to the Leopold Café firing at the people sat in and around the café, before moving on to the Hotel Taj Palace and Towers. The three remaining pairs went to the Trident Oberoi Hotels, Nariman House Jewish Community Centre (Chhabad House) and the Taj Group of Hotels. The attacks with firearms and explosives continued for the next 60 hours.
>
> 'The Sir Jamshetjee Jejeebhoy Group of Hospitals (JJGH) ... was closest to the site of the attack and received the majority of the casualties.'
>
> 'A total of 271 casualties arrived in the receiving area, of which 108 were brought in dead and taken to the mortuary for autopsy. Twenty-three patients received primary care as an outpatient. There were 140 patients (113 men and 27 women patients) who required admission to hospital, 13 of whom had minor injuries and were discharged the next day after appropriate treatment. Most patients were in the age group 20–39 year. The dominant injury pattern was limb trauma (seen in 117 patients); 12 patients had a neck injury. Most of the surviving patients had bullet injuries, followed by pellet injuries and a smaller number with blast injuries.'
>
> 'Surgical intervention was required for 127 patients, six of whom died after surgery. Abdominal trauma with visceral injuries was seen in 22 patients who required laparotomy for bowel and mesenteric injuries. There was a wide spectrum of injuries, including mesenteric tears and bowel perforations, as well as solid organ injury to the liver, spleen and kidneys. Chest trauma was seen in 29 patients, with haemopneumothorax being the main finding (14 patients). Most surgical procedures were for soft tissue or orthopaedic injury. Where possible they were done under local or regional anaesthesia, and included 30 primary wound closures, more than 40 debridements and six secondary wound closures. The mean duration of hospital stay was 12 (range 1–118) days.'
>
> 'The attack produced both bomb and bullet injuries in civilians, following the initial open-air mass attack at CST, Leopold Café and the three hotels. This contrasts with the train blasts in Mumbai in 2006 and London in 2005, which included only bombing. This was reflected in the injury patterns in surviving patients, which were predominantly bullet injuries, as bombing was more lethal.'
>
> (Extracts from Bhandarwar et al., 2012)

services for such incidents. These would be mobilized where possible. It is important that these doctors are not taken from the areas where they are needed in hospital, so reducing the hospital's capability to manage.

13.6 Recovery

Once the casualties have all been recovered and the scene made safe, the response phase moves to the recovery phase. This is usually managed by the local authority and command and control of the incident is handed from the police to the local authorities.

The recovery phase and returning a community to some normality is a huge operation that takes many months and in some cases many years. Comparing the recovery of the London bombings in 2005 with that of the Japanese tsunami in 2011: while they both go through the same recovery processes, the time frame is considerably elongated in Japan because of the infrastructure devastation caused by the tsunami. The radiation emitted from the Fukushima Nuclear Power Station effectively leaves the area abandoned for many years to come.

The resources mobilized in the response phase are undertaken to save life and the cost of the response is generally considered after the event. However, in the recovery phase the costs are considered before recovery is enacted, so economic pressures can, in some instances, have a bearing on the timeliness of a complete recovery.

The recovery phase is not just about rebuilding the homes and the infrastructure; it is also about repairing a community and the physical and mental welfare of those involved. This can take many years to complete. The victims of the Hillsborough disaster in 1989 fought for many years for justice for those they lost. It was not until 2012 that the case was reopened and in 2016 the coroner returned the verdict. For these families the emotional and mental impacts of the events have never healed (Chapter 9, this volume).

Organizations must also consider the welfare of their staff after the event. Many will have worked long hours and some will be affected emotional and psychologically by the events they have witnessed and experienced. It takes time and commitment for an organization to recover. It is not just about the logistics of ensuring the equipment and supplies have been replaced. Incident commanders must consider the welfare of their staff not only during the incident, but equally after it.

For health care services the recovery process is just as long. The cancelled elective operations and outpatient appointments means extending operating theatre lists, to ensure those patients who were deferred are given the treatments that they require. This is in addition to those patients who were directly affected by the incident, some of whom will go through many operations on the road to recovery, followed by physiotherapy, outpatient appointments and in some cases psychological counselling. The psychological impact on a community can last for years after the event.

One of the main parts of the recovery phase is that of identifying and learning the lessons from the incident and the response. It is important that the detail of the incident, the response and the decisions made are captured as soon as possible after the event. The legal frameworks that follow such an event will have commenced at the start of the incident, but it is in the recovery phase that pace gathers rapidly (Chapter 12, this volume).

Wherever a sudden death has occurred in the UK, the coroner will investigate to determine the cause of death and prevent a reoccurrence where possible. The Health and Safety Executive will look to see if there is liability on the part of a company or individual, the police will look at all the evidence to determine if there is a criminal case to answer and civil lawyers will look to see if there is a case to make, particularly against a public service body. There may well be a public enquiry (House of Commons Library, 2010). The evidential gathering of statements, log books, control voice recordings, etc. can go on for months. Any legal challenge brought against an organization results in many months of gathering evidence before any case is brought to trial.

Learning the lessons from an incident is one of the most difficult areas. Actions can be undertaken and measures put in place to prevent reoccurrence, but the real learning is with the people who experienced it and lived through it. Learning is achieved more through doing something and experiencing it, than it is through theoretical application. The latter helps, but it does not replace the actual experience.

> **Key Answers**
>
> - A mass casualty event is described by the UK NHS as 'a disastrous single or simultaneous event(s) or other circumstances where the normal major incident response of several NHS organisations must be augmented by extraordinary measures in order to maintain an effective, suitable and sustainable response'. This is further explained by the UK Department of Health as 'an event that creates casualties in the hundreds and not just in the tens'.
> - Simplistically, the difference between a major incident and a mass casualty incident is that of numbers affected. Those affected by the mass casualty incident will be in the hundreds and require arrangements in excess of usual major incident processes. A major incident may not have any casualties.
> - The challenges of responding to a mass casualty incident lie with the ability to safely manage and effectively care for those requiring medical care. With hundreds of people injured it is likely to overwhelm the ability of the local health economy to respond effectively. Therefore, plans need to consider managing casualties longer at the scene and distributing the injured much further afield than normal.

References

Advanced Life Support Group (2011) Major Incident Medical Management and Support. Available at: www.alsg.org/uk/MIMMS (accessed 11 December 2015).

Bhandarwar, A.H., Bakhshi, G.D., Tayade, M.B., Borisa, A.D., Thadeshwar, N.R. and Gandhi, S.S. (2012) Surgical response to the 2008 Mumbai terror attack. *British Journal of Surgery* 99, 368–372.

House of Commons Library (2010) Shipping: safety on the River Thames and the Marchioness disaster. Available at: www.parliament.uk/briefing-papers/sn00769.pdf (accessed 11 December 2015).

International Federation of Red Cross and Red Crescent Societies (2010) World Disasters Report 2010. Focus on urban risk. Available at: www.ifrc.org/Global/Publications/disasters/WDR/wdr2010/WDR2010-full.pdf (accessed 1 March 2016).

Lockey, D.J., MacKenzie, R., Redhead, J., Wise, D., Harris, T., Weaver, A., et al. (2005) London bombings July 2005: the immediate pre-hospital medical response. *Resuscitation* 66(2), ix–xii.

World Health Organization (2011) Disaster Risk Management for Health Fact Sheet: Mass Casualty Management. Available at: www.who.int/hac/events/drm_fact_sheet_mass_casualty_management.pdf (accessed 1 March 2016).

World Health Organization, Country Office for Nepal (2015) Situation Report # 19. Nepal Earthquake 2015. Available at: www.searo.who.int/entity/emergencies/crises/nepal/who-sitrep19-26-may-2015.pdf?ua=1 (accessed 27 November 2015).

Further Reading

Cabinet Office (2013) Emergency Response and Recovery. Non statutory guidance accompanying the Civil Contingencies Act 2004. Available at: www.gov.uk/government/uploads/system/uploads/attachment_data/file/253488/Emergency_Response_and_Recovery_5th_edition_October_2013.pdf (accessed 1 March 2016).

Department for Culture, Media and Sport (2008) *The Guide to Safety at Sports Grounds*. Her Majesty's Stationery Office, Norwich, UK.

Department of Health (2007) Mass Casualties Incidents – A Framework for Planning. Available at: http://webarchive.nationalarchives.gov.uk/20130107105354/http:/www.dh.gov.uk/prod_consum_dh/groups/dh_digitalassets/@dh/@en/documents/digitalasset/dh_073397.pdf (accessed 13 February 2016).

Events Industry Forum (2015) *Purple Guide to Health, Safety and Welfare at Music and Other Events*. Available at: www.thepurpleguide.co.uk (accessed 27 November 2015).

Home Office (2015) Joint Emergency Services Interoperability Programme. Available at: www.jesip.org.uk (accessed 27 November 2015).

NHS Commissioning Board (2015) Emergency Preparedness, Resilience and Response (EPRR). Available at: www.england.nhs.uk/ourwork/eprr (accessed 11 December 2015).

World Health Organization (2004) *World Health Report*. Available at: www.who.int/whr/2004/media_centre/en/index.html (accessed 27 November 2015).

World Health Organization and UK Health Protection Agency (2011) Disaster Risk Management for Health. Mass Casualty Management. Available at: www.who.int/hac/events/drm_fact_sheet_mass_casualty_management.pdf (accessed 27 November 2015).

14 Preparedness and Response to Pandemics and Other Infectious Disease Emergencies

J.S. Nguyen-Van-Tam[1] and P.M.P. Penttinen[2]

[1]*Professor of Health Protection, University of Nottingham School of Medicine, Nottingham, UK*
[2]*Head of Disease Programme: Influenza and Other Respiratory Viruses, European Centre for Disease Prevention and Control (ECDC), Solna, Sweden*

Key Questions

- What is the significance of pandemic influenza as a health emergency?
- What is the risk of another influenza pandemic and how quickly will it spread?
- What are the key principles underpinning emergency response to pandemic influenza?
- What main interventions are possible to reduce the impact of a future pandemic?
- How does the emergency response to pandemic influenza map across to other infectious disease emergencies?

14.1 Introduction

In this chapter we discuss the principles and practice of emergency preparedness and response as applied to infectious disease emergencies. The most significant of these in terms of size, likelihood and overall impact is undoubtedly a global influenza pandemic. Influenza is therefore used throughout this chapter as an exemplar; arguably, it is also the area of health emergencies in which preparedness and response activities have become the most sophisticated and highly developed.

14.2 Pandemic and Seasonal Influenza

Influenza is a well-recognized disease of humans, caused by infection with influenza viruses of types A and B. Influenza A viruses are further subdivided into subtypes denoted by 'H' and 'N', these being the major surface antigens of the virus: haemagglutinin (H) and neuraminidase (N), of which there are respectively 18 and 11 different variants currently identified in nature.

Both influenza A and B produce an acute illness most reliably recognized and distinguished clinically from other respiratory viruses by the triad of cough, fever and acute onset. Many other symptoms also occur quite commonly such as sore throat, myalgia, anorexia and headache; and fever is not always present, especially in the elderly. Thus, in clinical practice, it is not always possible to distinguish influenza definitively from other respiratory virus infections without laboratory testing. The likelihood of a correct clinical diagnosis increases when the clinician is informed about the circulation of the virus in the community through surveillance (see below).

Although influenza can produce a highly virulent illness, with profound myalgia and lethargy that send the sufferer to bed for several days, it is an 'old wives' tale' that says all influenza infections are this severe. In fact, the spectrum of illness in confirmed infections is very varied: 50–75% of infections are asymptomatic, while a small minority of cases progress rapidly to severe virus infection leading to pneumonia, hospitalization and death. Influenza infection also exacerbates pre-existing conditions such as cardiac failure, asthma, chronic bronchitis and diabetes; and some influenza-related hospital admissions and deaths occur that are never attributed to influenza because diagnostic tests are never undertaken. In very young children, the elderly and people of all ages with high-risk underlying illnesses such as heart disease, lung disease and diabetes, influenza is more likely to be a severe illness, with complications such as bacterial pneumonia, hospitalization and death. The public health burden of seasonal influenza therefore falls on the very young, the elderly and people of all ages with high-risk conditions and explains why such groups are targeted for routine seasonal vaccination.

Influenza viruses, especially influenza A, are highly variable and have a very high spontaneous mutation rate as they multiply inside the infected host. Via this mechanism, they ensure that human immunity due to having had the infection and recovered, or through vaccination, is relatively short lived. Each year the circulating influenza A and B viruses tend to be slightly different from the year before. Essentially this explains why influenza epidemics occur at some point during most winter seasons, why large parts of the population remain at least partially vulnerable to infection, and why influenza vaccines need to be reformulated and given to high-risk patients on an annual basis.

Pandemic influenza should not be seen as a separate disease entity from seasonal influenza. A pandemic is caused when a new type A influenza virus emerges that is not slightly different from previous circulating viruses (as in the year-on-year mutations seen with seasonal influenza; known as 'antigenic drift'), but rather *substantially* different ('antigenic shift'). When this happens and other preconditions are met (see Box 14.1), a pandemic virus is said to have emerged. In general terms the word 'pandemic' is often used for a disease affecting more than one continent, and the word 'epidemic' for more localized outbreaks of disease. Influenza type B is associated with the occurrence of epidemics, but not pandemics.

> **Box 14.1.** Preconditions for pandemic influenza
>
> - Novel influenza A virus, substantially different from current or recently circulating seasonal strains.
> - No or almost no pre-existing immunity in all or the majority of age groups.
> - Virus capable of producing clinically apparent illness in humans (symptomatic influenza).
> - Efficient person-to-person spread.

14.3 Epidemiology of Pandemic Influenza

Unlike seasonal influenza, which occurs annually with 'winter' epidemics in temperate zones, pandemics are rare and unpredictable events, with a roughly 3% per annum risk (based crudely on having had three pandemics in the last century). The occurrence of influenza pandemics prior to the 20th century is somewhat speculative, but from 1580 onwards there are well-described global outbreaks of febrile respiratory illness assumed to be influenza; and from 1889 onwards there is firmer sero-epidemiological evidence that such occurrences were actually influenza, although the virus was not identified by scientists until 1933. It nevertheless seems likely that human pandemics occurred in 1889/90 (probably H2) and 1898/1900 (probably H3); and in the 20th century pandemics were documented with complete certainty in 1918/19 (H1N1), 1957/58 (H2N2) and 1968/69 (H3N2); followed by the first pandemic of the 21st century in 2009/10 (H1N1).

A pandemic virus seems capable of emerging at any time of year, and there are often two to three separate pandemic waves each of typically 10–12 weeks, spread over an 18-month to 2-year period. It also seems to be the case that pandemic waves that coincide with 'winter' seasons are larger and peak higher than 'out of season' waves; for example, in both 1918 and 2009, early spring waves in the northern hemisphere were less extensive than the subsequent autumn second waves. This is probably explained by more favourable environmental conditions for influenza transmission in the autumn and winter months. Likewise, the largest pandemic wave in Australia occurred from June to September 2009 (its winter months) when in relative terms activity fell back during the corresponding warmer summer months in the northern hemisphere.

In the modern era, when international travel is extremely common, associated with a globally connected economy, pandemic viruses spread extremely rapidly around the world. For example, the 2009 H1N1 pandemic virus was identified in late April 2009, but within 6 weeks had spread to at least 60 other countries; in contrast, the 1918 pandemic virus took 18 months to spread worldwide. Since most of the population is, by definition, susceptible to a pandemic virus, the proportion of the population who become ill with symptoms (known as the 'clinical attack rate' or CAR) is typically high, in the region of 30–40% over the entire pandemic period. At least the same proportion again will be infected without getting symptoms (known as 'asymptomatic cases').

Both seasonal and pandemic virus infections cause extra deaths (known as 'excess mortality'). However, it is over-simplistic to say that mortality caused by a pandemic virus is always more substantial than for seasonal influenza. First, although this would be true for the very severe 1918/19 pandemic during which 2.5% of persons with symptomatic illness died (known as the 'case fatality ratio' or CFR), the CFR was far less extreme in 1957 and 1968 (<0.5%); and in 2009/10 the global CFR has been estimated at 0.02%, which is very much on a par with seasonal influenza. Second, for seasonal influenza the burden of deaths tends to fall mainly in the elderly (aged ≥65 years); but a consistent feature of pandemics is that the age-specific burden of deaths shifts, to a variable degree depending on the circulating strain, towards the younger age groups. This was especially apparent in 1918/19 when total mortality was massive (20–40 million deaths worldwide) and very large numbers of young adults died. One consequence of this shift is that pandemics have a far larger impact than seasonal influenza, even for the same absolute number of deaths, because deaths in younger people mean that the total years of life lost is far higher. Estimates for the USA suggest that in 1918/19 and 2009/10 (at opposite extremes of the severity spectrum) 64 million and 2 million years of life were lost respectively, compared with 600,000 in a typical inter-pandemic winter season. But of course, the counterbalance is that seasonal epidemics are annual events, whereas pandemics are sporadic; thus taken over time seasonal influenza may cause more total loss of life than pandemic periods. From an emergency planning perspective this does not alter the fact that pandemics are far harder to cope with because of their sheer intensity and size.

Seasonal and pandemic influenza share a number of epidemiological and clinical features (essentially the same symptoms, the same vulnerability in high-risk patient groups; excess mortality; seasonal variation) and rather unsurprisingly they are also interrelated in a virological sense. Although a pandemic virus is, by definition, novel when it first emerges and it tends to displace the seasonal influenza viruses that were in immediate prior circulation, it does not vacate this ecological niche by 'magically' disappearing at the end of the pandemic period. Instead, the pandemic virus establishes itself as a seasonal virus and continues to circulate. For example, the 1918 H1N1 pandemic virus remained in circulation through related but constantly evolving seasonal H1N1 viruses until 1957, when it was displaced entirely by the new H2N2 pandemic virus. 'Distant relatives' of the novel H3N2 pandemic virus that emerged in 1968 still circulate today as seasonal H3N2 viruses. And the 2009 H1N1 pandemic virus now circulates alongside H3N2 as a seasonal virus, having totally displaced the previous seasonal H1N1, which to all intents and purposes died out in 2008.

14.4 Detecting and Tracking Influenza Viruses

Clearly, in a situation where many different respiratory viruses co-circulate each winter, health professionals and health care managers need timely information on influenza activity and the earliest possible warning about epidemics to aid individual diagnosis. It is also important that pandemic viruses can be detected and tracked geographically and temporally.

Surveillance requires tracking of influenza in multiple dimensions. This begins with surveillance in the animal kingdom because influenza is in fact a disease where the natural reservoir is wild aquatic shorebirds. At least 16 of the 18 haemagglutinins that exist in nature and nine of 11 neuraminidases have been described in birds. In addition, there is evidence that the last four human pandemics have been due to viruses that contained avian influenza components. The recent outbreaks of avian influenza in birds related to subtypes H7N7, H5N1 and H7N9, all of which have spilled over into humans causing severe illness in a small number of people, also reflect the ongoing pandemic threats posed by avian viruses. At the time of writing, H7N9 is perhaps considered the highest known pandemic risk because there is evidence from China that alongside over 780 confirmed human cases to the end of May 2016, it is estimated that many thousands of asymptomatic cases have also occurred. This gives the virus multiple opportunities to adapt to spread more easily between humans although it has not yet done so.

Pigs are also considered alongside birds as a rather important species. Although several mammalian species can be infected with influenza there is some evidence that the pig, which can be co-infected with human and avian viruses, may have been an important 'virus mixing vessel' and implicated in the emergence of the pandemics in 1957, 1968 and 2009. Thus surveillance of influenza in the animal kingdom is important for human health, especially in aquatic birds, poultry flocks and swine herds.

Surveillance of influenza in humans also needs to be ongoing. This needs to be undertaken at two levels. First, virological surveillance is needed to characterize the viruses recovered from humans with influenza infection. This helps determine if the virus is changing and whether new variants or subtypes are emerging; if the vaccines in use are a good match for circulating wild strains; and if viruses are becoming resistant to anti-influenza drugs. One of the first signs of a new pandemic virus in humans is likely to be the detection of a novel virus in one of the laboratories that is part of a worldwide network of National Influenza Centres coordinated by the World Health Organization (WHO). Second, clinical surveillance is needed to monitor geographic and temporal patterns of illness in the population due to influenza. This can be achieved at household level by Internet surveys, and by monitoring health service contacts such as consultations with primary care doctors for influenza-like illness and hospital admissions and deaths ascribed to influenza, preferably linked to some degree of virological testing.

Strong surveillance systems are essential for accurate and timely decision making during public health emergencies. Syndromic systems for influenza, which aim to capture household morbidity, clinical illness, severe illness or mortality, can also be highly valuable during other respiratory disease outbreaks or environmental disasters. Whichever surveillance systems are in place in a country (and this is highly variable depending upon resources), pandemic planners must know which data are available and how to access, interpret and use these in real time to help manage an emerging crisis.

Seasonal and pandemic influenza are global phenomena; therefore, surveillance data also need to be shared internationally, to aid with the risk assessment and decision-making processes. The Global Influenza Surveillance and Response System (GISRS) of the WHO and the European Influenza Surveillance Network (EISN) are examples of mechanisms for sharing and analysis of essential surveillance data.

A systematic and rapid dynamic risk assessment process is needed to support timely decision making. This risk assessment process should assess the likelihood of novel emerging and zoonotic influenza strains seeding a pandemic, and also evidence-based scenarios for the evolution of the pandemic at national and local levels. The WHO, the US Centers for Disease Control and Prevention (CDC), the European Centre for Disease Prevention and Control (ECDC) and many national authorities have established methodologies for risk assessment.

14.5 Principles of Pandemic Preparedness and Response

Although pandemics are almost completely unpredictable and can vary greatly in their severity, some underpinning principles of preparedness and response can nevertheless be established.

The primary impact of a pandemic will be on health and health care services; there will be potential surge and a need for increased capacity at all points in the patient pathway, through illness onset to recovery or death. These are summarized in Fig. 14.1. But besides health care impacts, a severe pandemic also has the potential to impact on wider societal functions through: disruptions to essential services, supplies and markets (particularly food and fuel distribution); school closures (due to sickness absence, or possibly as part of a deliberate pandemic response strategy); staff absenteeism; business travel and tourism; public panic; and threats to social order. These wider impacts were not generally seen during the 2009 pandemic, but a more severe pandemic would be an altogether different experience.

The interventions applied to pandemic response can be broadly subdivided into public health measures, pharmaceutical interventions and the management of wider societal impacts. Due to influenza affecting large parts of the population, public health interventions rapidly become highly visible political issues and there is much pressure on public health authorities to react in a timely fashion. For example, the decisions during 2009 to do entry screening at airports for influenza were influenced by the need to react rapidly with a limited evidence base on the effectiveness of such a measure.

Unfortunately, many of the public health and pharmaceutical interventions still have a limited evidence base or specific unresolved issues about effectiveness. Therefore, the concept of using multiple interventions in a combined approach has been applied to overcome these issues; this is also known as 'layered containment', 'layered mitigation' or 'defence in depth', adopting the principle that multiple, partially effective, interventions are more likely to mitigate impact when applied simultaneously or in sequence.

- Increased physical facilities
- Increased patient throughput
- Maintaining adequate staffing
- Protecting staff at work
- Essential medicines
- Specialized facilities (intensive care)
- Antiviral drug supplies
- Vaccines for patients and staff
- Other essential clinical services (e.g. other medical emergencies, trauma and obstetric care, emergency surgery)
- Disposal of the deceased

(Nested circles, from outer to inner: More health care supplies and medicines; More staff; Increased health care capacity; Large numbers of people require health care at the same time)

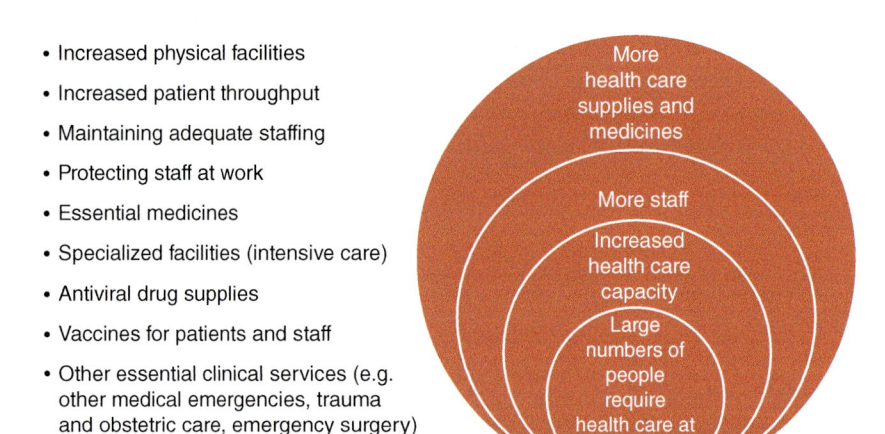

Fig. 14.1. Main health care surge impacts and challenges during pandemic influenza.

14.6 Public Health Measures

Although 'containment' is still a widely used term, it is also a misnomer; it is now widely accepted that pandemic influenza cannot be contained with currently available public health measures. Its rapid global spread is inevitable. The aim of public health countermeasures is therefore to reduce transmission. Although the primary stated aim is to reduce overall numbers of cases and deaths, more realistic secondary goals are to 'flatten' the epidemic curve and delay its peak. This reduces the peak demand for surge capacity (albeit at a potential cost of prolonging the duration of the response effort), i.e. buys time for additional preparation before that peak is reached; and potentially nudges the epidemic towards the time frame in which vaccines will be available (Fig. 14.2).

Public health countermeasures can be subdivided into:

- those with relatively strong evidence of effectiveness;
- those with unproven (uncertain) effectiveness; and
- those where there is evidence that the effect will be weak or minimal.

These are summarized in Table 14.1.

Some public health measures will have a differential effect according to the timing of their implementation. For example, school closures would have a larger impact if enacted at the start of a pandemic, whereas hand washing would have a smaller impact but be equally useful throughout the pandemic. Some measures are also harder to implement than others; school closures may have negative impacts on parental availability to work (including health care worker availability) and rational decisions have to be made about when schools should reopen. In contrast, hand washing requires supplies of soap and water (relatively easy to arrange) and sustained public compliance over many months (relatively hard to enact).

Just as all medicines have adverse effects, public health interventions can have potential negative consequences, even if they are considered effective. For example, a mass vaccination programme will divert resources from other public health priorities or social distancing measures. Because of this the modern concept of operations suggests a proportionate approach is most appropriate, whereby those measures that are most difficult to implement, or have the most dramatic negative consequences, should be reserved for severe pandemics. Whichever public health measures are considered, planners need to understand and rehearse the practical steps that would be required to enact them and to monitor and manage the negative consequences.

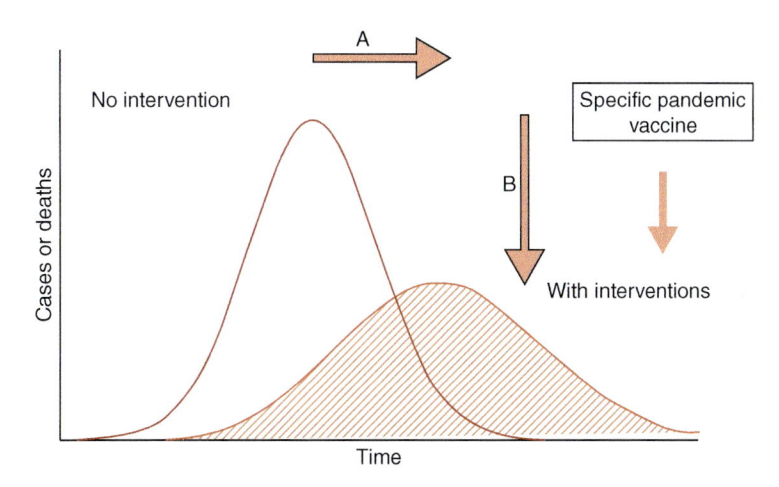

Fig. 14.2. Principles of public health countermeasure effectiveness (A: epidemic curve peak is delayed, B: peak surge value is reduced in height). (From Nicoll, A. and Lopez Chavarrias, V. (2013) National and international public health measures. In: Van-Tam, J. and Sellwood, C. (eds) *Pandemic Influenza*, 2nd Edn. CAB International, Wallingford, UK.)

Table 14.1. Summary of public health countermeasure effectiveness.

Minimal	Unknown/uncertain	Some effect
Travel advisories	Respiratory etiquette (presumed)	Regular systematic hand washing
Entry screening	Mask wearing by public (across multiple settings)	Mask wearing by infected symptomatic persons
Border closures (unless complete)	Early self-isolation (presumed)	Reactive school closures
Internal travel restrictions (minor delaying effect?)	Quarantine	Proactive school closures (better)
	Workplace closures	
	Home working/reduced meetings	
	Cancellation of mass gatherings	

14.7 Pharmaceutical Measures

Pharmaceutical measures can be subdivided into antiviral drugs, vaccines and antibiotics.

14.7.1 Antiviral drugs

Neuraminidase inhibitors (NAIs) are specific anti-influenza drugs that first became available in 1999 and the main two drugs in this class, oseltamivir and zanamivir, were deployed during the 2009 pandemic. Data from clinical trials suggest that in mild cases of seasonal influenza in mainly healthy adults, NAIs reduced symptoms and complications of influenza such as hospitalization. In contrast, observational data from the 2009 pandemic suggest that mortality was reduced overall by 19% when NAIs were used in patients hospitalized with pandemic influenza; and by 50% if treatment could be started within 2 days of symptom onset. There are also data available that suggest that NAIs successfully reduce mortality in patients who are severely ill with avian influenza, which may be important if the next human pandemic emerges from an avian source, e.g. H7N9. Overall, these data suggest that NAIs are an important pharmaceutical response measure, especially in severely ill or rapidly deteriorating patients, and especially considering that vaccines will arrive relatively late in most countries, as was the case in 2009/10.

Many developed countries, the WHO and multinational private companies have decided to stockpile NAIs for pandemic response for the foreseeable future. A stockpile is a large investment in preparedness and must be managed and refreshed when it expires. Some countries have attempted a 'rolling stockpile', analogous in mechanism to a strategic petroleum reserve, where the stockpile is used to support everyday medical needs and continuously refreshed; however, this mechanism is difficult to sustain if seasonal use of NAIs is limited. Once activated, usage of the stockpile must be controlled so that the national supply is not exhausted prematurely, for example half-way through the pandemic. An important component of preparedness planning is that that practical mechanisms need to be worked out and rehearsed which allow the correct target patient groups timely access to treatment; this is a complex exercise involving many levels of the health care system and difficult to achieve in practice.

14.7.2 Vaccines

The challenge for seasonal and pandemic influenza vaccine production is that the vaccine needs to be developed against a specific influenza strain or strains. Therefore, the WHO decides twice per year, 6 months in advance, which strains should be included in the northern and southern hemisphere seasonal vaccines. When a pandemic is declared, based on the WHO declaration, vaccine companies halt the production of seasonal vaccines and start developing and producing the pandemic vaccine.

Although small quantities of pandemic H3N2 vaccine were available in the USA late on during the 1968/69 pandemic, the first widespread availability of pandemic vaccines was during the 2009 pandemic. The virus emerged in April 2009 yet the first vaccine supplies arrived in early October 2009, reflecting a current minimum production lead time of 5 months. Further delays can be caused by inadequate planning of the distribution and mass administration of the vaccine. Due to overwhelming demand and finite production capacities, during the 2009/10 pandemic only countries with pre-existing purchase agreements or access to WHO donations were able to access the pandemic vaccine.

Analyses of effectiveness suggest that these vaccines were 70–80% effective in averting cases in vaccinated populations. These two facts illustrate the present paradox related to pandemic influenza vaccines. They are undoubtedly an effective preventive measure at individual level, yet their overall public health effect is blunted by their late availability in epidemiological terms (Fig. 14.3) since pandemic vaccine production cannot begin until after the pandemic virus has emerged. Thus new production mechanisms, new vaccines or novel mechanisms for preparedness are needed that will facilitate more rapid access to large-scale volumes of vaccine.

It is possible to acquire stockpiles of specific vaccines in advance of a pandemic (known as 'pre-pandemic vaccine'). This has been done in some countries for H5N1, reflecting ongoing concerns that this virus presents a high pandemic threat. If an H5N1 pandemic were to emerge in the future, countries with a stockpile would have a vaccine ready to deploy at the very start of the

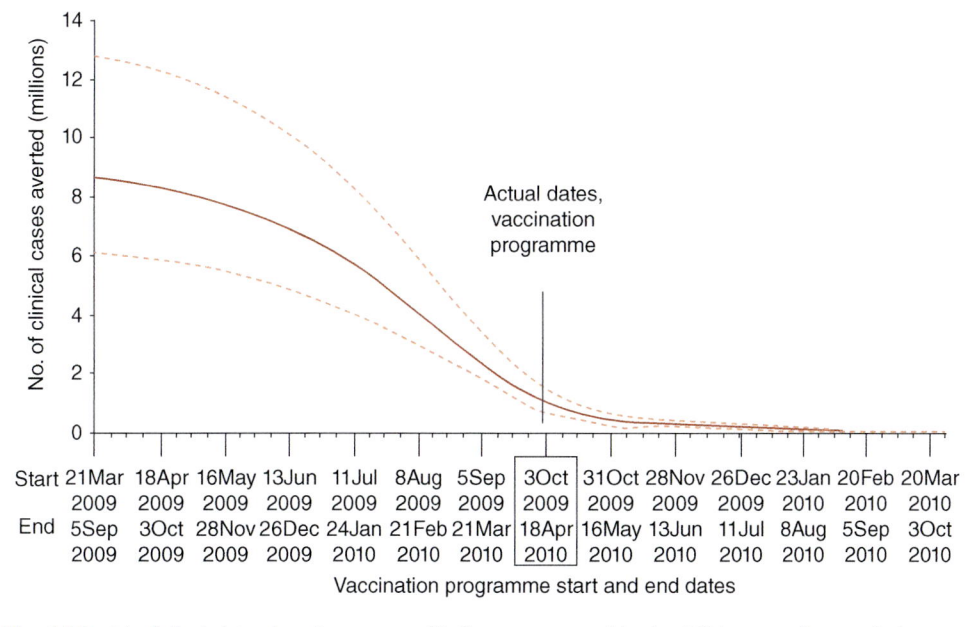

Fig. 14.3. Modelled data showing cases of influenza averted in the USA according to timing of availability of pandemic vaccine (from Borse, R.H., Shrestha, S.S., Fiore, A.E., Atkins, C.Y., Singleton, J.A., Furlow, C., et al. (2013) Effects of vaccine program against pandemic influenza A(H1N1) pdm09, United States, 2009–2010. *Emerging Infectious Diseases* 19, 439–448. Available at: http://dx.doi.org/10.3201/eid1903.120394 (accessed 17 June 2016).

crisis, perhaps to protect health care workers and the most vulnerable patients. This is a gamble however, and if an H5N1 pandemic does not emerge, the vaccine will have no practical use (it will not be effective against other influenza viruses) and the stockpile will expire, unused.

14.7.3 Antibiotics

One of the commonest complications of influenza is secondary bacterial respiratory infection including pneumonia. This suggests that having a rolling stockpile of antibiotics available for health service use is an effective preparedness measure also against influenza pandemics. In addition, it has been suggested that pneumococcal vaccines may have a role to play in reducing pandemic-related pneumonia, but this is unproven and the concept is not well developed.

14.8 Communication and Ethical Issues

In common with many health emergencies (Chapter 8, this volume), effective communication with health care workers and the public is essential and has not always been well handled in the past. Particular problems relate to pandemic influenza because the crisis is relatively long lived (up to three epidemic waves over 18–24 months) compared with 'big-bang' incidents, which are over in a few days or weeks. Information at the beginning of a pandemic (soon after the emergence of the virus and before a pandemic is formally declared) is a known problem; epidemiological data are scarce, the CFR is uncertain and estimates may change or evolve before they become stable. During the 2009 pandemic many public health authorities perceived that communicating the level of uncertainty was the key communication challenge faced. It is difficult to communicate prevention and treatment strategies related to antiviral drugs and stress the importance of vaccines while being honest about the latter's likely late arrival. Sustaining public interest and maintaining public compliance with infection control and hygiene advice over a sustained period is also difficult. The advent of the Internet and social media provides an easily accessible platform where divergent views, often challenging authority and scientific evidence, can be heard. This makes risk communication a highly complex and difficult operating environment and changes the way that public authorities need to communicate.

Severe pandemics also raise specific ethical issues related to the scarcity of health care, medicines and vaccines in relation to potentially overwhelming demand and other competing health care priorities. Difficult decisions may need to be taken that will disadvantage some people in favour of the greater good. And health care workers may refuse to attend work if the (perceived or actual) level of personal danger is high. While there are no definitive solutions, planning for these ethical dilemmas and the development of an a priori ethical framework is advantageous.

14.9 Other Pandemic Threats and Infectious Disease Crises

Preparedness against pandemic influenza demonstrates various aspects of generic health preparedness, as influenza pandemics are global, with a local impact; they evolve rapidly, while also lasting over several years; they infect large proportions of the population, while smaller proportions, but still large numbers of cases, experience severe disease or death. Therefore, it is a good model to consider for all infectious diseases in preparedness planning.

As many of the elements for health sector preparedness are the same across diseases or conditions (e.g. command and control, surveillance, surge capacity in health care, risk communication, social distancing), being well prepared against influenza pandemics offers some surety against being able to respond effectively to other infectious disease emergencies.

The increasing interconnectedness of the global population enhances the spread of infectious diseases. The emergence of severe acute respiratory syndrome coronavirus (SARS) in China in 2003 and its subsequent rapid global spread within weeks suggest a pandemic 'near miss' occurred; given the absence of specific antiviral treatments, the absence of a vaccine and the high CFR, this would have been serious if it had taken off. The introduction of SARS into a hospital in Toronto is an example in how an infectious disease can rapidly paralyse the health care system in a modern, developed city and cause large economic losses to the country; and illustrates how rapid and coordinated public health actions can limit such impact. Similarly, although there is emerging evidence that the source of sporadic cases of Middle Eastern respiratory syndrome coronavirus (MERS-CoV) appears to be contact with dromedary camels, and human-to-human transmission occurs mainly in hospitals and is unsustained, its potential pandemic threat cannot be ignored.

The successful eradication of smallpox in the 1970s and the subsequent end of the global vaccination programme has left an increasingly large portion of the global population susceptible to this highly infectious and virulent disease. The causative virus currently exists only in freezers of a handful of highly secure laboratories monitored by the WHO. However, many countries continue to stockpile smallpox vaccines mainly as a biodefence measure against an intentional release of the disease. Many countries have invested in large amounts of money (often eclipsing spending on health preparedness) in biodefence measures against intentional release of infectious or toxic biological agents.

Increasingly complex and geographically dispersed food production and distribution chains are increasing the threat posed by contamination of food products during some part of the chain. Recent years have seen, for example, multi-country outbreaks of hepatitis A linked to frozen berries and various salmonella outbreaks related to internationally distributed eggs or meat products.

The Western Africa Ebola outbreak that commenced in March 2014 took the world's public health professionals by surprise. While the disease was known to cause localized outbreaks and much publicity had surrounded it in prior years, including popular movies, most professionals believed it could be contained by rapid actions within small, localized, outbreaks. The slowness of the local, regional and global response meant that within 6 months the event had become the most important communicable disease threat in Africa since smallpox was eradicated. The lack of investment by the national and international authorities meant that inadequate surveillance and response capacities were in place, a vaccine was not available, and even basic supportive treatment and isolation could not be made rapidly available for many patients. This event highlights the difficulty of ensuring sufficient resources are available for planning and preparedness outside an ongoing crisis, even with a highly visible and understandable threat profile.

Notwithstanding, most developed countries in stable geographical and political environments rank the threat of pandemic influenza as the top civil societal threat and thus have existing, extensive, pandemic preparedness plans. Indeed, when asked in January 2015 at the World Economic Forum to list the top infectious disease threats for the years ahead, Dr Tom Frieden, Director of the US CDC, said: 'I think we'd all start with flu'.

Key Answers

- Based on potential size of impact and likelihood of occurrence combined, most national and international authorities regard a future influenza pandemic as the most significant health emergency threat.
- The risk of another influenza pandemic cannot be accurately quantified and its timing cannot be predicted; however, it is widely accepted that, in the modern era, spread from its place of emergence will be very rapid, with extensive global distribution after approximately 2–3 months.
- It is widely accepted that pandemic influenza cannot be contained at source. The overriding aim is to mitigate impact. The established principle is that multiple, partially effective, interventions are more likely to mitigate impact when applied simultaneously or in sequence.
- The main types of interventions will be public health measures; health care surge capacity; pharmaceutical measures (antiviral drugs and antibiotics); and vaccines. The last three depend heavily on resources and funding; and, based on current technologies, vaccines will not be available until 4–6 months after a pandemic virus has emerged.
- Pandemic influenza is seen as a good template for preparedness against other infectious disease emergencies. It is a global phenomenon, with local impact; it evolves and spreads rapidly, yet can last several years; it infects a large proportion of the population, with the potential to produce large numbers of severe cases or deaths.

Further Reading

Borse, R.H., Shrestha, S.S., Fiore, A.E., Atkins, C.Y., Singleton, J.A., Furlow, C., et al. (2013) Effects of vaccine program against pandemic influenza A(H1N1) pdm09, United States, 2009–2010. *Emerging Infectious Diseases* 19, 439–448. Available at: http://dx.doi.org/10.3201/eid1903.120394 (accessed 17 June 2016).

European Centre for Disease Prevention and Control (2014) Overview of EU Member States' national influenza pandemic preparedness plans. http://ecdc.europa.eu/en/healthtopics/

pandemic_preparedness/national_pandemic_preparedness_plans/Pages/influenza_pandemic_preparedness_plans.aspx (accessed 14 February 2016).

Gates, B. (2015) The next epidemic – lessons from Ebola. *New England Journal of Medicine* 372, 1381–1384.

Hashim, A., Jean-Gilles, L., Hegermann-Lindencrone, M., Shaw, I., Brown, C. and Nguyen-Van-Tam, J. (2012) Did pandemic preparedness aid the response to pandemic (H1N1)2009? A qualitative analysis in seven countries within the WHO European Region. *Journal of Infection and Public Health* 5, 286–296.

Miller, M.A., Viboud, C., Balinska, M. and Simonsen, L. (2009) The signature features of influenza pandemics – implications for policy. *New England Journal of Medicine* 360, 2595–2598.

Napoli, C., Fabiani, M., Rizzo, C., Barral, M., Oxford, J., Cohen, J., et al. (2015) Assessment of human influenza pandemic scenarios in Europe. *Euro Surveillance* 20(7), 29–38.

Nguyen-Van-Tam, J.S. and Bresee, J. (2013) Pandemic preparedness and response. In: Webster, R.G., Monto, A.S., Braciale, T.J. and Lamb, R.A. (eds) *Textbook of Influenza*, 2nd edn. Wiley, Chichester, UK, pp. 453–469.

Nicoll, A., Brown, C., Karcher, F., Penttinen, P., Hegermann-Lindencrone, M., Villanueva, S., et al. (2012) Developing pandemic preparedness in Europe in the 21st century: experience, evolution and next steps. *Bulletin of the World Health Organization* 90, 311–317.

Van-Tam, J. and Sellwood, C. (eds) (2012) *Pandemic Influenza*, 2nd edn. CAB International, Wallingford, UK.

World Health Organization (2011) Strengthening Response to Pandemics and Other Public-Health Emergencies. Available at: www.who.int/ihr/publications/RC_report/en/index.html (accessed 14 February 2016).

World Health Organization (2013) Pandemic Influenza Risk Management: WHO Interim Guidance. Available at: www.who.int/influenza/preparedness/pandemic/influenza_risk_management/en/ (accessed 14 February 2016).

15 CBRN Incidents

R.P. Chilcott[1] and S.M. Wyke[2]

[1]*Head of Toxicology, Department of Pharmacy, University of Hertfordshire, Hatfield, UK*
[2]*Principal Public Health Scientist, Centre for Radiation, Chemical and Environmental Hazards, Public Health England, Chilton, UK*

Key Questions

- What is a CBRN incident?
- What are the common causes?
- What are the key response elements for dealing with casualties during a CBRN incident?
- Are there specific considerations for fatality management?

15.1 Introduction

The threat from global terrorism has increased over the last decade. In particular, it has become apparent that certain organizations may wish to seek injury or death to unprotected civilians through the deliberate release of hazardous substances such as chemical warfare agents or toxic industrial chemicals. Indeed, such events have already occurred, most notably during the Tokyo subway attack using sarin, a volatile nerve agent. Less well publicized were several previous attacks using the persistent nerve agent VX and more recently the polonium-210 incident in London.

While still relatively uncommon, mass poisonings have highlighted the need to ensure that first responders have both the training to recognize incidents and the available resources to mitigate the health effects of exposure to toxic materials. The potential impact of such events has led many governments to review existing response arrangements and to develop, where necessary, new and improved means of dealing with major incidents. Recently, the European Parliament and the Council of the European Union (EU) adopted new legislation that aims to improve the coordinated response to cross-border health threats (Decision 1082/2013/EU). It sets provisions on notification, ad hoc monitoring and coordination of public health measures in response to serious cross-border threats to health from biological, chemical and environmental events, as well as events that have an unknown origin. The legal instrument applies to all EU Member States and is comparable to the International Health Regulations in its content, requirements and adoption of a multiple-hazards approach. The purpose of this chapter is to review some of the causes of both accidental and deliberate release, general principles of identification, contact tracing, initial operational response and fatalities.

15.2 Causes of Accidental and Deliberate Release

Incidents or releases can be accidental, via deliberate intent or as a result of natural disasters (Table 15.1). They may be on a small or large scale, and can give rise to a number of primary or secondary chemical casualties and fatalities. The International Federation of the Red Cross and Red Crescent Societies estimates that between 1998 and 2007 there were 3200 incidents involving chemical releases with approximately 100,000 people killed and nearly 1.5 million affected. The number of casualties following a release depends on the location and type of incident and can range from a few to thousands.

In the 1970s and 1980s governments focused their effort on the scientific basis for chemical safety to strengthen national capabilities, including safety of production, storage and transport of chemicals. However, high-profile chemical incidents such as the Seveso disaster in Italy in 1976, the 1981 toxic oil incident in Spain and the 1984 Bhopal explosion in India led to increased recognition of the public health impact of chemical incidents. The deliberate use of biological and chemical agents against civilian populations has further highlighted the threat from terrorist and paramilitary organizations and, perhaps most importantly, has identified areas in which improvements to emergency response arrangements can be made. Indeed, the UK has recently implemented a revised emergency response procedure for managing mass casualty incidents based on new scientific evidence.

Large-scale incidents are rare, but when they do occur there is a risk that resources may be stretched or overwhelmed in the affected countries. Within the EU, expert help may be requested from neighbouring nations to deal with the incident effectively.

It is prudent to plan for the response to a mass emergency involving toxic chemicals, even though such events are rare. Nevertheless, such an eventuality may develop at a rate and reach a magnitude sufficient to impose a major crisis on society. A well-developed mechanism exists at the European level where humanitarian aid and civil protection assistance can be requested by EU and non-EU countries in response to disasters.

Following the advent of severe acute respiratory syndrome, avian flu, 9/11 and other events, authorities began to realize that the spectre

Table 15.1. Examples of major international chemical, biological and radiation incidents involving mass casualties or worried well.

Year	Location	Description of incident	Impact
1976	Seveso, Italy	Industrial accident: airborne release of dioxin from industrial plant	No immediate human deaths 3300 animal deaths 80,000 animals slaughtered Chloracne in approximately 200 individuals and led to some of the highest body doses of TCDD ever measured
1981	Spain	Deliberate: ingestion of an oil fraudulently sold as olive oil, resulting in TOS	300 deaths 20,000 people affected Led to chronic illness
1984	Bhopal, India	Industrial accident: MIC leak from a tank	3800 immediate deaths 15,000 to 20,000 premature deaths 500,000 exposed to the gas
1995	Tokyo, Japan	Deliberate release of a chemical warfare agent (sarin)	12 deaths 2500 casualties 500 homes uninhabitable
2000	Enschede, The Netherlands	Industrial accident: explosion at a fireworks factory	20 deaths 562 casualties Hundreds of homes destroyed 2000 evacuated
2001	USA	Deliberate: contamination of postal items with anthrax	Five deaths 11 individuals contracted cutaneous anthrax 31 tested positive for exposure Thousands required antibiotic prophylaxis
2005	Hemel Hempsted, England	Industrial accident: explosions at oil storage facility	No deaths 2000 people evacuated
2006	Scotland	Accidental (environmental): anthrax from contaminated animal skin	One death (inhalational anthrax) 73 'contacts' required antibiotic prophylaxis
2006	London, England	Deliberate: polonium-210 poisoning of one individual	One death Thousands of worried well Multiple locations contaminated
2008	London, England	Accidental (environmental): anthrax	One death (inhalational anthrax) <15 contacts required antibiotic prophylaxis
2009	Hungary	Industrial accident: toxic mud	Ten deaths 286 injured persons (121 required treatment in hospital) Major environmental and economic impacts
2009	China	Melamine in milk	Six deaths 50,000 children hospitalized 300,000 infants affected

TOS, toxic oil syndrome; MIC, methyl isocyanate; TCDD, 2,3,7,8-tetrachlorodibenzo-*p*-dioxin.

of hazards that might seriously affect societies are many-fold and that predicting them is difficult and further complicated by the need to deal with different threats. It was therefore deemed reasonable to adopt a 'generic' or 'all-hazards' approach to improve the overall preparedness and response capacity of authorities. An all-hazards approach allows for better planning and preparation for situations where more than one type of agent may potentially be released or where the agent is unknown. This approach is also more likely to facilitate and enable responsible authorities to deal with more complex public health incidents and emergencies potentially involving more than one type of hazard, such as the Icelandic volcanic ash cloud in 2010 and the effects of climate change. It is also important to bear in mind that there are unique chemical risks associated with most non-chemical disasters. For example, when a structure is damaged by flood or an earthquake, chemical substances are also spilled and mixed with other chemicals which may pose a risk to first responders or to people returning to their homes after the event. Accidents initiated by natural hazards or disasters that result in the release of hazardous materials are commonly referred to as 'Natech' or 'na-tech' accidents.

15.3 General Principles of Identification and Contact Tracing

15.3.1 Incident recognition

The starting point for any incident response is the realization that an event has actually occurred. The public health effects of exposure to chemical, biological or radiological materials can only begin to be mitigated when response plans are activated following incident recognition. While this seems like an obvious statement, it is worth considering that most biological and radiological agents have a latent period of at least several hours. Furthermore, many chemicals do not provoke immediate signs or symptoms of intoxication and will silently produce pathological changes in the absence of overt clinical features or other warning signs such as unusual or pungent odours. Classic examples of such chemicals include phosgene and sulfur mustard. A further consideration is that the onset of clinical effects following exposure to chemicals that act predominantly via dermal absorption (such as the nerve agent VX) may also be subject to a latent period which, in turn, may be affected by other factors such as the anatomical location of the exposure and the prevailing environmental conditions.

A small number of chemicals have specific signs and symptoms of exposure (toxidromes) which may alert responders to an incident. For example, chemicals that act via inhibition of cholinesterase (such as nerve agents) may produce overt signs of poisoning consistent with nicotinic or muscarinic stimulation such as miosis ('pin point pupils') and excessive salivation. For this reason, several countries have developed algorithms to assist in the recognition of exposure to key threat agents but it must be noted that only a relatively small group of chemicals have such characteristic toxidromes; there are literally thousands of chemicals that could potentially be used in a deliberate release incident and their acute effects are non-specific such as coughing, headache, nausea and dizziness.

Overall, incident recognition is the 'starting pistol' for responding to a major incident and requires first responders to be well trained and vigilant.

15.3.2 Contact tracing

Major incidents can have a significant and long-lasting impact on the physical and mental health of those involved and may particularly affect individuals who are more susceptible to adverse circumstances such as children, the elderly and pregnant women. Following a major incident, it may be necessary to establish a register of affected individuals in order to:

- provide advice on relevant immediate and longer-term public health interventions that may be required;
- provide reassurance to the public that their care is paramount;
- reassure the worried well to avoid them overwhelming local services; and
- facilitate epidemiological investigations.

The purpose of establishing a register is to identify the population affected by or exposed to the incident so that:

- appropriate advice on relevant immediate interventions can be provided;
- access to appropriate services can be facilitated;
- reassurance can be provided to the public;
- health impact assessments of the incident can be initiated; and
- the longer-term health implications of the incident can be investigated.

Those affected by or exposed to an incident can often disperse quite rapidly immediately afterwards, therefore mechanisms need to be in place to enable health registers to be set up quickly. Health registers are key to facilitating communication between relevant services and to affected individuals, as well as enabling emergency and health services to better assess the health impact and identify longer-term health implications of an incident.

15.4 Initial Operational Response

Although radiological and biological agents are clearly of concern, chemical exposures will often require more rapid clinical intervention to mitigate potential health effects. Obvious examples of chemicals associated with a rapid onset of intoxication include nerve agents and hydrogen cyanide, both of which may be lethal within minutes. In contrast, there may be a potential therapeutic window of at least several hours or more for the effective administration of medical countermeasures against biological or radiological contaminants. Thus, chemical exposure presents a different chronological challenge from incidents involving radiological or biological materials and, for this reason, chemical incident response timescales should be considered as the denominator for planning an all-hazards response.

Historically, military doctrine has been used as a basis for developing civilian operational responses to CBRN (chemical, biological, radiological and nuclear) incidents. However, there are many considerable and necessary dichotomies between military and civilian preparedness. For example, military personnel are generally healthy and highly trained individuals who may carry appropriate personal protective equipment (PPE) and medical countermeasures on their person. In contrast, there will necessarily be a delay between initial exposure and on-scene arrival of appropriate equipment, countermeasures and trained emergency response personnel for civilian incidents. Thus, while some military practices can be applied to civilian incidents, the two are generally incongruous. Moreover, it cannot be assumed that all civilian casualties will be able to undertake self-evacuation, disrobing and/or decontamination. Indeed, a number of susceptible populations have been identified for whom a functional-needs approach will be required.

Many countries have developed plans for responding to mass casualty CBRN incidents from which a number of common features can be distilled and are discussed below. However, it cannot be overemphasized that actual events may require incident managers to consider ad hoc changes to long-standing plans and have the presence of mind not to delay activities critical to reducing potential health impacts by doggedly adhering to written procedures.

15.4.1 Evacuation

The primary response to any incident must be to evacuate individuals from the source of exposure. For ambulant casualties, this should involve rapid self-extraction from the source of contamination (commonly referred to as the 'hot zone' or 'red zone') to a relatively safe location (the 'warm zone' or 'white zone'), which ideally should be upwind, at an elevated and safe distance. Such an apparently simple step can present a number of practical problems due to inherent uncertainties regarding the source, location and magnitude of the contaminant and environmental factors such as changes in wind direction. Any difficulty in establishing a safe distance from the point of release should lead to consideration of alternative tactics, such as advising 'shelter in place'. Such issues may be location-specific and so only resolvable at the time of an incident.

Non-ambulant casualties may arise through traumatic injury related to the incident or pre-existing disability. Life-threatening injuries may necessitate stabilization of the patient prior to movement. However, evacuation would be a priority over stabilization if the hot zone were to be overtly life-threatening. The evacuation of non-ambulant casualties from the hot zone may require a 'snatch rescue' for which appropriate protective equipment would be required to prevent the rescuer from becoming a casualty.

In the presence of an airborne hazard, it would be inappropriate for a responder to attempt a snatch rescue without some level of respiratory protection (e.g. an 'escape hood'). As the primary role of the emergency services is to save lives, this could pose a considerable dilemma, especially where non-ambulant hot-zone casualties are visibly distressed.

Historically, the clinical management of casualties has traditionally been performed after evacuation and decontamination from the hot zone. The arrival of specialist assets to the scene would likely cause a delay in evacuation and decontamination and may consequently reduce the survivability of a hazardous material incident, especially for non-ambulant, high-priority patients. As such, several countries have developed a capability to allow advanced clinical care within hot-zone environments. In the UK for example, the ambulance services can deploy a Hazardous Area Response Team (HART; England and Wales) or Special Operations Response Team (SORT; Scotland) to perform potentially life-saving procedures such as endotracheal intubation, intra-osseous antidote administration and haemostatic interventions.

15.4.2 Disrobing

The simple act of removing clothing (disrobing) is a highly effective method for removing external contaminants from casualties and should be implemented at the very earliest opportunity during an incident response. It is often stated that disrobing can remove up to 90% of contamination from an individual, although there does not appear to be any scientific evidence for this claim. It is likely that that the figure of 90% is derived from the 'rule of nines' on the assumption that all areas of the body (except the hands and face) are covered in relatively impermeable clothing. However, the act of disrobing is undoubtedly a highly effective and practical means of reducing exposure to hazardous materials providing some basic precautions are taken. For example, clothing should ideally be cut from the body to limit spreading and/or inhalation of the contaminant.

The effectiveness of disrobing is time-critical, as contaminants may diffuse through the fabric layers to the skin surface. For a single layer of cotton clothing, the amount of liquid contaminants (such as chemical warfare agents) that can be removed by disrobing decreases substantially during the first 30 min of exposure. Therefore, it is essential that disrobing be performed as soon as practically possible, ideally within 10–20 min of exposure. Perhaps most importantly, disrobing should always be performed before full body decontamination, as water can transfer contaminants through clothing on to the skin surface (Fig. 15.1).

There are a number of obvious practical challenges associated with disrobing, not least maintaining the privacy and dignity of casualties and the provision of replacement garments. Where a formal re-robe provision is unavailable, alternatives such as clothing from a local retailer should be considered. In extremis, blankets, foil sheets or even opaque plastic sheets

CBRN Incidents

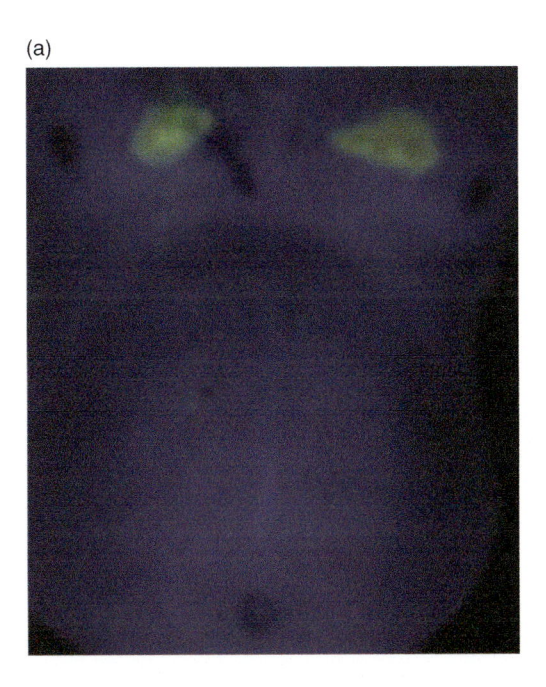

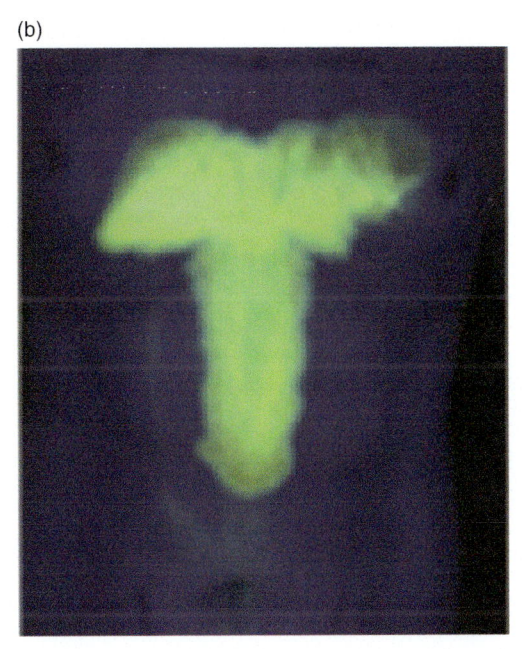

Fig. 15.1. Residual skin contamination (indicated by light areas) on torso of volunteer following water shower decontamination. Removal of clothing before showering results in efficient removal of contaminant (a). In contrast, there is an increased spread and intensity of residual skin surface contamination when clothing is worn during showering (b). (Reproduced with permission of the US Biomedical Advanced Research and Development Authority © 2013.)

(e.g. bin liners) may offer a temporary re-robing capacity. Potentially contaminated clothing should be placed in plastic bags and be readily identifiable and stored in a safe and secure manner so that personal effects may be returned or used as evidence in subsequent criminal investigations.

15.4.3 Decontamination

Decontamination can be defined as the process of removing hazardous material(s) both on or available to the external surfaces of the body in order to reduce local or systemic exposure to a contaminant and thus minimize the risk of subsequent adverse health effects. In the USA, mass casualty decontamination is commonly achieved using the 'ladder pipe' method (Fig. 15.2), which showers individuals with large volumes of water mist. In contrast, the UK has a bespoke decontamination capability that uses heated water within a temporary, sheltered structure (Fig. 15.3).

Decontamination is generally performed in the warm zone following evacuation of casualties from the hot zone. However, some countries such as Israel have developed a system for directly transporting casualties from the scene of an incident to medical centres with decontamination facilities. This negates delays associated with deploying temporary decontamination structures to the scene of an incident.

It is important to consider decontamination as part of a single procedure that extends the initial process of disrobing rather than a separate stage of casualty management. As with disrobing, the effectiveness of

Fig. 15.2. Demonstration of the ladder pipe system (LPS) for mass casualty decontamination. Individuals are directed to walk through a high-volume water mist generated by overhead and side fogging nozzles from two adjacent fire tenders. (Reproduced with permission of the US Biomedical Advanced Research and Development Authority © 2013.)

Fig. 15.3. Mass casualty decontamination unit ('MD1') deployed by UK specialist responders. Photograph acquired during an exercise shows a group of ten individuals in high-visibility ponchos (from 'disrobe packs') waiting to enter the unit in accordance with a 'traffic light' system (inset, top left). A schematic of the unit (inset, top right) indicates position of disrobe area, two side corridors (each with five shower areas ('S') for decontamination of ambulant casualties) and a central corridor (used by responders to observe or instruct individuals) which can be adapted for processing non-ambulant casualties. Air heaters and a boiler for shower water are at the rear of the tent and so not shown in this image. (Reproduced with permission of Public Health England © 2013.)

decontamination is also time-critical and so should also be performed as soon as reasonably possible.

In general, most current decontamination plans involve the use of water as a decontamination medium. Such 'wet' processes rely on the availability of water (optionally containing detergents or other excipients such as bleach) to wash and rinse potentially contaminated areas of hair and skin. In most developed countries, water is likely to be readily available. However, wet decontamination has several potential disadvantages:

- Waste effluent may increase the mobility of a contaminant within the environment.
- Viscous substances may be difficult to remove.
- Oily substances may have limited dissolution in water and so detergents may be required.
- Water may enhance the spread of contaminants over the skin surface.
- Showering may potentially lead to hypothermia where heated water or shelter is unavailable.

- Some studies have indicated that water may enhance the dermal absorption of certain contaminants. This effect can be markedly reduced by limiting the duration of wet decontamination to less than 90 s.

Bleach (hypochlorite) has been suggested as a means of neutralizing chemical contaminants and animal studies have confirmed some degree of effectiveness. However, the threshold dose of bleach for eye irritation (0.5%) is of questionable value for achieving rapid and complete neutralization of chemicals on the skin surface.

Evidence-based recommendations for optimizing aqueous decontamination have recently become available (Table 15.2) and are currently being implemented by the UK emergency services.

Physical cleaning of the skin surface (e.g. through the use of a flannel or facecloth) can improve the effectiveness of decontamination by c.20%. Where self-cleaning is not part of a decontamination procedure, the active stage of decontamination occurs after showering when the skin and hair are dried (e.g. with a towel). In such cases, the act of drying is actually the key step for removing residual contamination. Therefore, caution must be exercised when handling materials that have been used to dry individuals after decontamination showering (as they are most likely to be contaminated).

The main disadvantage of wet decontamination is the time taken to deploy and set up a bespoke resource at the scene of an incident. As discussed earlier, disrobing and decontamination are time-critical and so any delay will necessarily reduce clinical benefit. Therefore, more rapid forms of decontamination are preferable during the early phase of an incident response. One potential alternative is 'dry' decontamination; the use of any available absorbent material to blot contaminants from the skin surface. In general, dry decontamination is based on the absorbent properties of powders or fabrics to passively draw contaminants from the skin surface and is particularly effective for liquid contaminants. Indeed, emergency responders in some French regions have adopted a process of dry decontamination (using fuller's earth or 'Gant Poudre') to achieve more rapid, initial decontamination prior to showering.

In practical terms, dry decontamination may be achieved using any available absorbent material. Examples include absorbent paper (kitchen, toilet and facial tissues), incontinence pads, nappies (diapers) or cotton fabrics (Fig. 15.4). However, caution would be required when handling or disposing of such products after use as a decontaminant.

Table 15.2. Summary of conditions for optimization of aqueous (shower-based) decontamination according to the 'ORCHIDS Protocol'.

Parameter	Optimal condition
Shower water temperature	35°C
Shower duration	60–90 s
Detergent	0.5% (v/v) Argos™ or FloraFree™
Washing aid	Cotton flannel (facecloth)

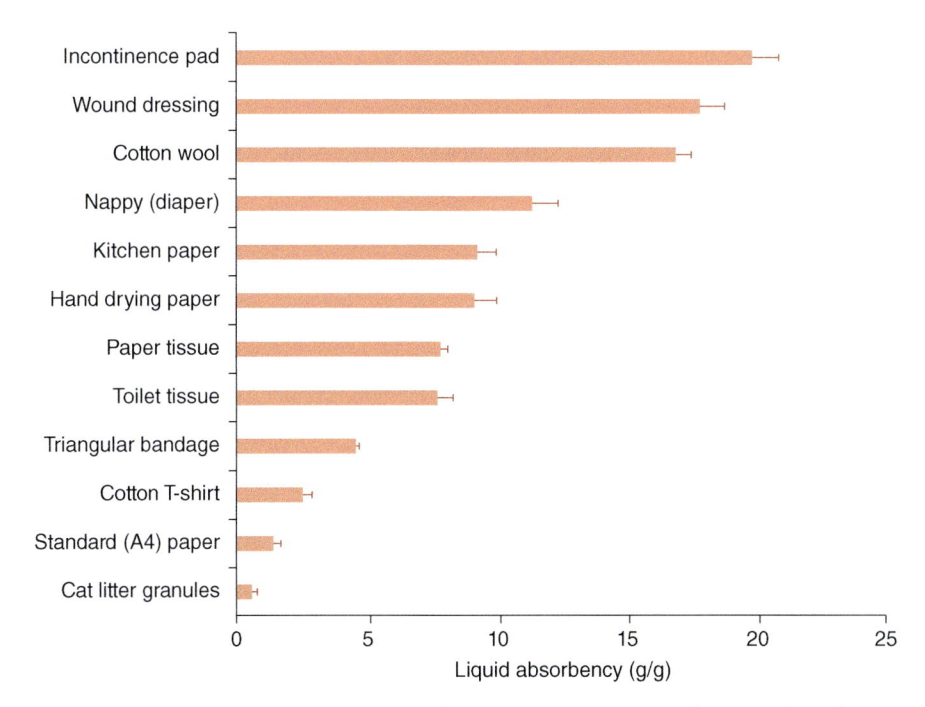

Fig. 15.4. Range of absorbencies (gram of oil absorbed per gram of test product) for materials that may be readily found in ambulances or domestic environment. Values are means with the standard deviation represented by error bars. Data provided for illustrative purposes only and do not represent an endorsement of any particular type of product.

15.5 Fatalities

The management of fatalities arising from a CBRN incident varies according to cultural beliefs and so there is no globally harmonized protocol; a fact that is reflected by current international guidance. However, it is possible to identify principal elements based on available guidance such as that developed by the UK's Home Office. The primary elements (Fig. 15.5) of fatality management include:

- *Identification of the causative agent*. This step provides the necessary information upon which to base subsequent risk assessments and selection of appropriate PPE for emergency responders.
- *Recovery of remains*. This would normally be overseen by a law enforcement agency in order to prevent loss of evidence. As this work would be performed in the hot zone, appropriate PPE would need to be worn by all personnel.
- *Decontamination*. Following transport from the hot zone, the remains would be subject to appropriate decontamination in the warm zone to prevent further spread of contamination and to reduce the hazard to emergency responders.

```
CBRN incident
      ↓
Identify causative      Hot zone
      ↓
   Recovery
      ↓
Decontamination        Warm zone
      ↓
 Identification
      ↓                 Cold zone
Temporary storage
      ↓
   Mortuary
      ↓                Public areas
Burial/cremation
```

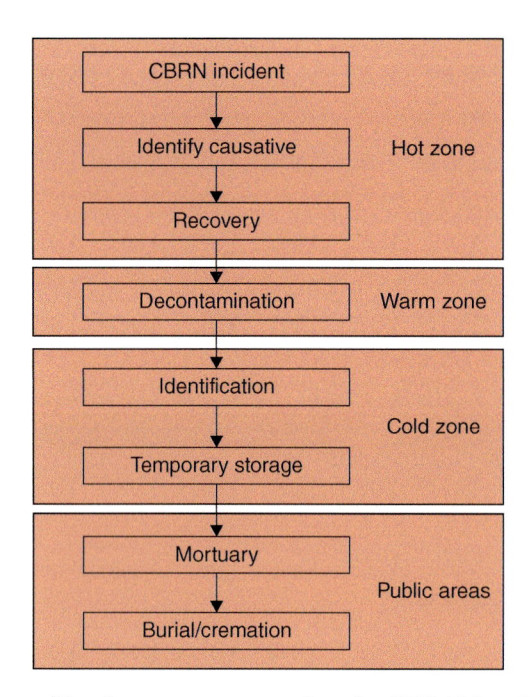

Fig. 15.5. Summary of fatality management actions for CBRN (chemical, biological, radiological and nuclear) incidents.

- *Temporary storage.* Normally, bodies would be placed in temporary storage close to the scene of the incident but at a safe distance ('cold zone'). This would allow for initial identification of fatalities (by forensic analysis and possibly visual identification by friends or family), preliminary pathological examination (if facilities permit) and further gathering of forensic evidence.
- *Mortuary.* When deemed safe, fatalities would be subsequently transported to a mortuary facility where a full autopsy could be performed using appropriate PPE along with a further opportunity for visual identification. Where appropriate, remains may be placed in sealed body bags.
- *Burial/cremation.* Considerations may include the need for extraordinary containment requirements (e.g. lead-lined or charcoal-filled coffin) and proximity of the grave or crematorium to environmentally sensitive areas such as underground water supplies (aquifiers), schools, etc.

It should be emphasized that both fatalities and survivors of a CBRN incident may be regarded by law enforcement agencies as evidence or witnesses, and so this will need to be factored into any response plans.

15.6 Practicalities for Health Care Response

Detailed response plans are generally written around the scene of an incident and downstream health care facilities such as casualty-receiving hospital(s). In this respect, the practicalities of handling potentially contaminated casualties (such as control of casualty movement, selection of appropriate PPE, decontamination facilities, etc.) are usually well defined. However, there are some circumstances under which health care staff may inadvertently or unexpectedly come into contact with potentially contaminated individuals; for example, emergency response teams who are first to arrive on scene or health care workers at non-emergency facilities (such as a general practice) dealing with self-reporting casualties. Part of the problem in dealing with such events is recognizing that a CBRN incident has occurred. In the UK, emergency response staff are trained to use the 'STEPS 1-2-3 PLUS' concept when approaching unconscious or incapacitated individuals, especially where there is no apparent cause:

- STEP 1 – single casualty – proceed as normal.
- STEP 2 – two casualties – approach with caution.
- STEP 3 – three or more casualties – consider a CBRN or HAZMAT (hazardous materials) incident.

It should be noted that, under the UK's new 'Initial Operational Response' (IOR) procedure, 'PLUS' refers to the instigation of evacuation, followed by immediate disrobe and decontamination by any available means. This replaces the previous 'Model Response' of withdrawing from the area to await the arrival of specialist response assets. However, the new process requires that any form of intervention be subject to a dynamic risk assessment to ensure the personal safety of first responders.

Self-presenting casualties at non-emergency health care facilities pose a significant problem, as front-line staff (such as receptionists) may not be trained to identify the signs of a CBRN incident. Under such circumstances, good communication is vital. In particular, casualties should be reassured that they can obtain the best treatment either at the scene of the incident or at the receiving hospital(s). Where possible, every effort should be made to ensure that potentially contaminated individuals remain outside the health care facility. If signs and symptoms of intoxication are apparent, the casualties should be encouraged to undertake immediate disrobe and decontamination.

15.7 Summary

Exposure of large numbers of individuals to hazardous chemical, biological and radiological materials may occur through a variety of means, including natural, accidental and deliberate (malicious) acts. Such extreme events are rare, but can be catastrophic both in terms of acute and chronic health effects.

Good planning is key to delivering an optimal emergency response. Such plans should include the means by which victims can be identified, traced and contacted for long-term evaluation. In terms of the initial phases of a major incident, the key components are incident recognition, evacuation, disrobe and decontamination, with protocols for ensuring the dignity of the deceased in alignment with cultural beliefs. Acquisition of forensic samples is a consistent element throughout all phases of a response. However, the health of casualties should be the primary concern, with adequate systems in place to ensure rapid disrobe and decontamination of potentially contaminated individuals guided by evidence-based practice.

> **Key Answers**
>
> - Events involving the accidental or deliberate release of extremely hazardous materials are referred to as CBRN (chemical, biological, radiological and nuclear) incidents. Such incidents may affect a substantial number of individuals and so are synonymous with mass casualty events.
> - Materials likely to be involved in a CBRN event include (but are not limited to) chemical and biological warfare agents, toxic industrial chemicals and medical radioisotopes.
> - Time is a critical factor when responding to CBRN incidents. Evacuation to a safe distance is the first priority and should be immediately followed by removal of contaminated clothing (disrobe) and decontamination by any available means.
> - It should be remembered that a CBRN incident is likely to be considered a criminal act. While it is important to respect cultural beliefs, the remains of the deceased may provide important forensic evidence and so be necessarily subject to atypical arrangements.

Acknowledgements

Work relating to the transfer of contaminants from clothing to skin during wet decontamination was funded in whole or in part with Federal funds from the Office of the Assistant Secretary for the Preparedness and Response, Biomedical Advanced Research and Development Authority. Work relating to the effectiveness of dry decontamination products was derived from independent research commissioned and funded by the Department of Health Policy Research Programme (EDICTAS 047/0200). The views expressed in this publication are those of the author(s) and not necessarily those of the Department of Health.

Further Reading

Baker, D. (2004) Civilian exposure to toxic agents: emergency medical response. *Prehospital and Disaster Medicine* 19, 174–178.

Chilcott, R.P. (2013) Initial management of mass casualty incidents. In: Arora, R. and Arora, P. (eds) *Disaster Management: Medical Preparedness, Response and Homeland Security*. CAB International, Wallingford, UK, pp. 311–324.

Chilcott, R.P. (2014) Managing mass casualties and decontamination. *Environment International* 72, 37–45.

Duarte-Davidson, R., Orford, R., Wyke, S., Griffiths, M., Amlôt, R. and Chilcott, R. (2014) Recent advances to address European Union Health Security from cross border chemical health threats. *Environment International* 72, 3–14.

European Commission (2005) Interim document: Technical guidance on generic preparedness planning for public health emergencies. Available at: http://ec.europa.eu/health/ph_threats/Bioterrorisme/keydo_bio_01_en.pdf (accessed on 31 January 2014).

Lake, W.A., Fedele, P.D. and Marshall, S.M. (2013) *Guidelines for Mass Casualty Decontamination during a Terrorist Chemical Agent Incident: Volumes I and II. ECBC-SP-036*. Edgewood Chemical Biological Center, Aberdeen Proving Ground, Maryland.

16 A Military Case Study

D. Ross and A. Charnick

Army Health Unit, Former Army Staff College, Surrey, UK

Key Questions

- What are the lessons that can be learned from military deployments to ensure a timely response and resilience?
- How does the military prepare, maintain and recover its forces in response to an emergency?
- What is the 'continuum of care'?

Box 16.1. Setting the scene

An East African country is heading towards instability as a result of organized groups of militias, which have formed together in an armed opposition to the elected government of the country. The militias have been moving across the country and a significant number of internally displaced persons (IDPs) are being driven into the government-controlled areas. This has affected the government's ability to combat the militias, as it has to split resources away from combat operations to support the IDPs. The government has requested assistance from both the African Union and the United Nations (UN) in dealing with the insurgency and the IDPs and their security and life support. Following a Security Council meeting it has been agreed that a UN-mandated peacekeeping force will be deployed to assist in the stabilization of the country and to provide support to the IDPs. The UK is providing an infantry battalion and logistic troops to the UN-led force. In addition to this, West Africa is in the middle of an outbreak of viral haemorrhagic fever.

16.1 Introduction to the Continuum of Care

The scenario presented in Box 16.1 above is not beyond the realms of possibility; the type of event described has occurred in the past and will no doubt occur again. The military will be required to do what it does well, and that is plan and deliver, while reviewing and adapting the plan continually. As the plan adapts, the health threats need to be reconsidered as they may change with the plan. Any military planning process will need to consider the 'what if' element of a situation and have contingency plans available for those possible reasonable events.

The UK Armed Forces consider that there are three phases to any deployment:

- pre-deployment;
- maintenance (operationally deployed); and
- recovery (post-deployment).

Anyone serving will therefore always find themselves in one of these phases. However, one might argue that if personnel are to be resilient and therefore able to undertake their role to the best of their ability, there are other factors that need to be considered before they enter the pre-deployment phase. In particular, factors that occur before an individual is recruited, during their selection and training are all pertinent. Similarly, once an individual leaves the armed forces there are lessons to be learned that might improve the resilience of their successors. Therefore, recently in considering resilience in medical staff and more widely armed forces personnel the concept of a *continuum of care* has been introduced (Fig. 16.1). Underpinning this must be high-quality data that are robustly analysed at all stages of the continuum. Using the scenario described above, some of the key tenants of this continuum are teased out in this chapter, including indications of how this could be applied to any team that is responding to an emergency situation.

16.2 Pre-deployment Phase

16.2.1 Medical intelligence

Prior to deploying forces to an operational theatre it is important to ensure that the troops are medically prepared and informed of the health threats they may be exposed to, once deployed. Failure to do so could (and probably would) have unwanted consequences in respect of available personnel. The threats are identified by access to *medical intelligence*: the product of processing medical, bioscientific, epidemiological, environmental and other information related to human or animal health and to identify those areas that are a potential threat to personnel.

Therefore, in preparation for the deployment, information is sought in respect of the hazards from the environment, disease and industrial activity within the country of operations. Within the scenario described, this includes the likelihood of viral haemorrhagic fever (VHF) spreading throughout Africa. Additionally, information relating to the location and quality of host nation medical facilities, and medical facilities within adjacent countries, is also gathered. As well as the health and medical information, data are sought on the availability and quality of water and food

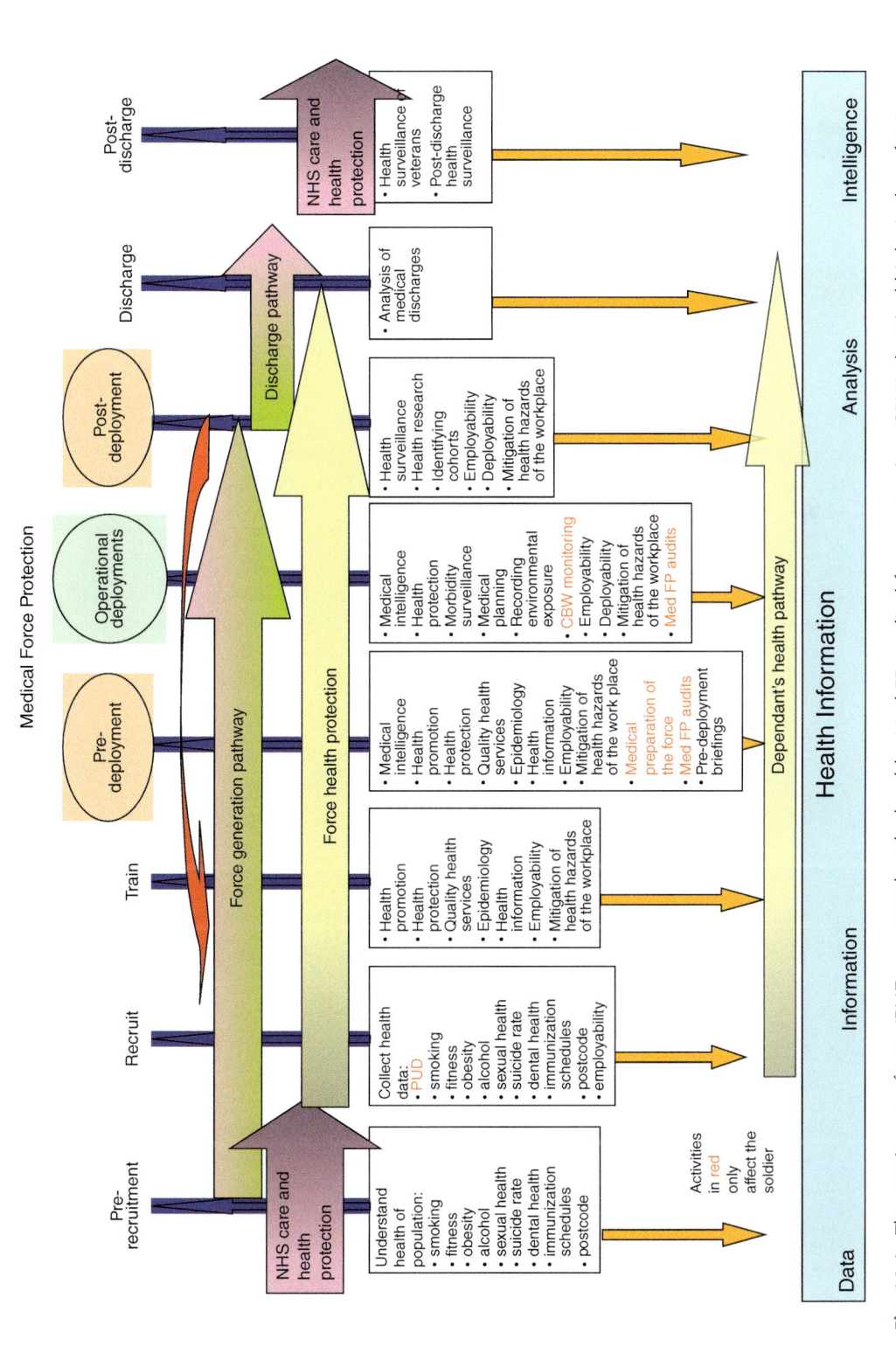

Fig. 16.1. The continuum of care (PUD, personnel undeployable; Med FP, medical force protection; CBW, chemical biological warfare). (Reproduced with permission from Ross, D.A. (2012) Preventive Medicine in the 21st Century – A Population Challenge. *J R Army Med Corps* 158(2), 77–78.)

supplies in the region of operation, to enable local procurement of fresh fruits and vegetables as well as meat products and potable water.

An initial reconnaissance party will be deployed to verify the open-source information that has been used is correct and appropriate for the area of deployment. All of the information is then analysed and a *medical force health protection* plan is developed in which appropriate and proportionate action is identified to mitigate those risks to an acceptable level so as to not affect the operation by undue manpower loss. The Allied Joint Medical Publication, *AJP-3.13 Allied Joint Doctrine for Force Protection*, defines force protection (FP) as:

> Measures and means to minimize the vulnerability of personnel, facilities, materiel, operations and activities from threats and hazards in order to preserve freedom of action and operational effectiveness thereby contributing to mission success.

Force health protection (FHP) as a subset of FP is the sum of all efforts to reduce or eliminate the incidence of disease and non-battle injuries (DNBIs) to enhance operational health readiness and combat effectiveness. The risk assessment will look to identify the risks to the operation and the individuals. For example, a major outbreak of gastrointestinal illness will potentially stop a small operation whereas an individual with a snake bite will not have an impact. Bricknell and Moore describe a 'health risk management' tool that may be used to undertake the analytical process required to produce the *medical force protection* (MFP) plan. Additionally, they describe how it may be useful to consider hazards in the following hierarchy:

- conventional battle hazards (e.g. bullet, bomb and blast);
- non-battle traumatic hazards (e.g. road traffic accidents, training accidents);
- infective hazards;
- chemical hazards;
- radioactive hazards;
- environmental hazards;
- psychological hazards; and
- ergonomic hazards.

16.2.2 Selection of personnel to deploy (force generation)

All personnel to be deployed must be fit to perform their role in the environment they are deploying to (i.e. 'fit for task'). Having screened out personnel during recruitment and training, one could make the assumption that all serving personnel are fit for role. However, this would be naive as illnesses may develop during service, some of which may mean an individual may not be able to adequately protect his/her own health if deployed. It is essential therefore that through a robust occupational health service, personnel be medically assessed for fitness to perform their role in the environment that they are required to deploy to; that is, undergo a medical risk assessment (MRA). In this scenario, for example, an individual who routinely takes antihypertensive medication might be considered to be a risk if he/she deploys where routine re-supply of medication may not be possible. UK Armed Forces personnel are normally categorized into three categories: (i) medically fit for all deployments (MFD); (ii) medically limited for deployments (MLD); and (iii) medically non-fit for deployments (MND). Anyone who is not MFD must always have a full MRA before he/she deploys.

16.2.3 Team health

Prior to deployment, an assessment of the preventive measures that should be offered to individuals based on the MFP plan will be undertaken. In conjunction with UK civilian authorities such as the Foreign and Commonwealth Office, the National Travel Health Network and Centre and TRAVAX, the plan will include the disease threats that may require pre-exposure vaccinations and/or the issuance of chemoprophylaxis (e.g. for malaria). Armed forces personnel are usually protected against a set of core diseases, which includes diphtheria, hepatitis A and B, polio, tetanus and yellow fever. This therefore minimizes the number of additional vaccines that may need to be offered. In addition to vaccinations, appropriate malaria chemoprophylaxis, if required, will be issued to individuals and a central plan to enable the impregnation of uniforms with insect repellents put in place. Timing of the deployment will be critical in determining what vaccines (and malaria chemoprophylaxis) can be offered. For example, a requirement to deploy in 14 days will not allow a full pre-exposure course of rabies vaccine to be offered if it is deemed an important preventive measure. Wherever possible, vaccination does not take place in the country except in extreme circumstances. The doctrine is to deploy a force that is medically protected from the outset.

A pre-deployment brief will be undertaken which will include a cultural brief to ensure that the risk of causing offence to the local population is reduced. Winning 'hearts and minds' with the local population will be vital. From a health perspective, personnel will be informed of food and water safety, bite prevention and any other risk to their health, and how to deal with the health threats. This briefing will also cover aspects of road safety, safe driving and industrial threats that may be encountered. It will also be the opportunity to raise the issue of mental health, which is key to resilience. All members should be informed that they will undoubtedly undergo a period of adjustment reaction and, depending on the things they see and do, may experience an acute stress reaction. It should be emphasized that the majority of personnel will recover from this stress reaction within a relatively short period of time. Only a few will go on to experience post-traumatic stress disorder (PTSD) (Chapter 9, this volume). Military chaplains play an important role in resilience awareness, training and management.

Families and loved ones are important contributors to the resilience of team members. Therefore, a welfare brief aimed at families as well as the troops should be given. This should include details about communication with their family member (e.g. use of phones, Internet access, etc.) while deployed and who is the rear party point of contact for the family. It should also cover what to do if a member of the family is ill. Additionally, deploying personnel will be encouraged to ensure that their partner/family is aware of where key documents are kept and also to make a will. Increasingly within the UK military community, individuals are being encouraged to bank a DNA sample in the event that their body needs to be identified at a later date. Inevitably this issue itself opens up an ethical discussion, with some individuals being concerned that the sample may be used for different purposes at a later date.

16.2.4 Equipment

Equipment relevant to the location such as mosquito nets, camp beds, insect repellent and sun cream will be issued if required. From the medical information there might also be a need to order appropriate pesticides/rodenticides and equipment to respond to the disease vectors present within the area of operation.

Individuals must ensure they take appropriate personal kit that can be carried easily to mitigate against ergonomic hazard. For the military this may be relatively straightforward; although recent experience shows that the ergonomic risk has increased, with individuals carrying increasing kit loads. For others this may not be the case and therefore organizations should endeavour to give clear instructions around personal kit.

Referring back to the scenario, in light of the VHF outbreak in the neighbouring country there will need to be further medical planning. This will need to consider the potential risk to personnel that will be deploying to support the United Nations (UN) force and whether additional protection or equipment is required such as enhanced personal protective equipment (PPE).

16.2.5 Training

Alongside the general pre-deployment brief, specific training to selected personnel in respect of sanitation, pest control, water supplies and its protection, communicable disease management and morbidity reporting will be required. It is important that any training given is validated before individuals are deemed fit to deploy. Usually training will be both individual and collective. In the military, the environmental health officers and technicians, supported by public health specialists, are an essential element in delivering this. One further consideration should be the employment and training of local personnel within the area of operation to undertake the routine tasks such as pest control and waste management. This will enable the more specialist staff to concentrate on their primary function without distraction of secondary time-consuming tasks.

16.3 Deployed Phase

Having selected the members of the team, once deployed many of the issues addressed during the pre-deployment phase will be re-rehearsed on arrival in the country. All personnel will need to undergo a period of acclimatization and during this period a further briefing will be undertaken, which will contain additional information as well as confirming that which they have already been told. There may also be a period of some confirmatory training to ensure that personnel are fit for task in country.

Figure 16.2 illustrates diagrammatically the issues that will need to be addressed both before and while deployed. Once in country there is a need to initiate appropriate vector control programmes to minimize the threats from mosquitoes, sandflies, etc. and to provide pest control measures against rodents and, if appropriate, feral dogs and cats. One aspect to reduce attraction as a food source or breeding site to these pests is to provide and manage effective sanitation and waste management systems and camp hygiene.

In addition, an effective system will need to be introduced to regularly monitor the potability of water supplies and a system to ensure the security and potability of water supplies from both field and established infrastructure. Water testing includes an in-theatre field testing system and reach back to an accredited laboratory.

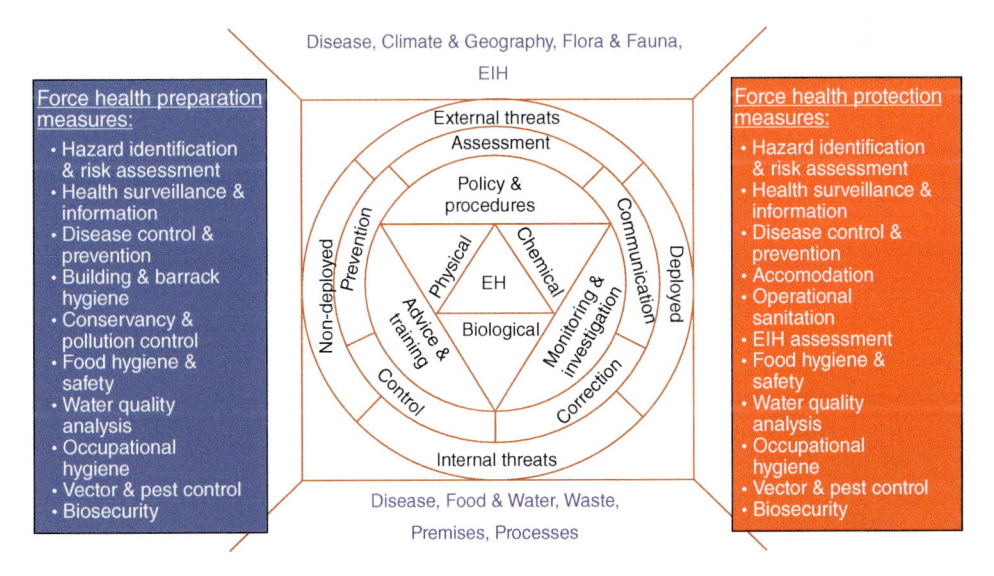

Fig. 16.2. Environmental health model for force health protection (EH, environmental health; EIH, environmental and industrial hazard).

There will be a number of challenges that the deployed force will face. To mitigate against some of these and ensure resilience in the force, there are several actions that must be carried out.

16.3.1 Daily routine and briefings

All individuals will go through some form of adjustment reaction depending on their level of experience. Such reactions will not necessarily occur at the same time for all. As well as being able to identify individuals who may experience such reactions, a daily routine helps mitigate the risk of such reactions turning into acute stress reactions. The two most important things to routine are set mealtimes and a regular daily brief cascaded through the components of the force.

16.3.2 Disease and non-battle injury surveillance

It is important to monitor the health of the deployed force through DNBI surveillance. Although we live in a technological age, a simple system that does not rely on electricity and/or large amounts of human resource is preferable. Such a system may be as simple as the RASIO rates (i.e. the rates of respiratory, alimentary, skin, infection and other diseases) through to a more complex notification system that includes all notifiable diseases of interest and common diseases of each anatomical/physiological body system (e.g. dental, gut, etc.). It is probably best to start with RASIO plus dental rates as these can be reported daily to commanders. Perhaps unexpectedly to many, history tells us that dental morbidity is one of the most common and costly reasons for reporting sick in a deployed environment. Reporting must be in the form of rates rather than numbers, and therefore an accurate count of the population at risk (PAR) will be required. Such rates must be reported at the command group's daily briefings, as this will enable commanders to make decisions as to what tasks can be carried out; for example, a sudden increase in respiratory cases such as influenza could have an impact on operational tasks.

16.3.3 Medical provision

During the pre-deployment phase an analysis of the medical support that would be required would have been made. Many solutions will have been predicated by availability of local host nation and UN facilities. Robust and locally available health care services may mean that the deploying force may only require access to a team medic with paramedical skills; whereas limited local support may dictate taking, at the minimum, a primary care physician. If local facilities are used there must be a process of assuring the care given to the deployed force. At its simplest level this will require someone to visit and view the facilities.

16.3.4 Casualty evacuation chain

In any deployment a proportion of individuals will become injured and/or unwell, which will mean they may need to be evacuated home. The evacuation route and resources must be planned and modified as circumstances change. For instance, the original plan may have included using civilian flights from the local airport for the walking wounded; however, if civilian flights stop flying because of real or perceived risk to their staff and aircraft, then the plan may need to be changed. In the scenario in this chapter, this may become a reality if the VHF outbreak starts spreading across Africa. In 2014 this was seen in real life when airlines stopped flying to Sierra Leone because of an outbreak of Ebola.

Depending on the role of an evacuated individual the organization may need to replace him/her. It is important therefore that there is a list of individuals in the UK who are fully prepared and ready to deploy at short notice. In the military this is termed *battle casualty replacement*.

16.3.5 Security

On arrival in the country it will be important to ensure that the deployed force is protected from external threats. These external threats will present as physical threats from insurgent forces, which may be realized by a direct or indirect assault on a location or vehicles and personnel in transit. There are also lower-level threats resulting from thefts and assaults and/or attempts to hack into computer and communications systems. Information obtained through a hack could be used to inform the planning of assaults, or present a threat to individuals' bank accounts, and in extreme but rare occasions could be realized in threats to families.

16.3.6 Political and NGO engagement

Early on arrival there will need to be engagement with the local government and in particular any non-government organizations (NGOs) that are delivering aid. Historically the relationship between NGOs and the military has had the potential for friction. The points of tension have been identified as: 'organisational structure and culture; tasks and ways of accomplishing them; definitions of success and time frames; abilities to exert influence and control information; and control of resources'. If the force is to be effective, clear roles for it must be defined and other organizations, be they international organizations, local government or NGOs, must be kept informed of the military role to avoid unnecessary misunderstanding and unrealistic expectations. Military commanders must also understand the role and aims of the other organizations within the theatre and what their capabilities and constraints are. It is beholden on both sides to understand each other to ensure success. For that to happen it is important that there is regular contact and coordination between all parties to ensure a common understanding, avoid gaps in provision of support, avoid duplication of effort and develop a sharing of information for the relief of those who are suffering.

16.3.7 Trauma risk management

The UK military for some time has used a system called *trauma risk management*. This allows individuals who may have been exposed to a traumatic situation such as involvement in a road traffic accident to discuss the situation with a trained non-medical peer, who will be able to assess the impact of the situation on that individual and signpost (if required) him/her to professional support. It is not critical stress debriefing, which has been shown to be ineffective in preventing PTSD and may cause more harm than good. Some other organizations including the UK Department for International Development have adopted this practice.

16.4 Post-deployment and Recovery Phase

This is the time that individual team members, particularly those not from a formed unit (often reservists), will be most vulnerable. It is important that there are measures in place to address reintegration into normal daily activity and relationships. Formed units will have the existing command structure, which in general should identify problems at an earlier stage. For individual augmentees there will need to be in place a system for both the organization to 'reach out' to individuals and for individuals to be able to 'reach in' should they need support.

16.4.1 Health problems

When an individual is first deployed he/she is likely to experience an adjustment reaction. Team members may find it difficult to talk about their experiences with their family members and feel a sense of loss from the team that they worked with. A few may go on to suffer from mental health problems, including substance misuse.

16.4.2 Long-term surveillance

Some health problems including mental health ones may not manifest themselves until years after the response. A system needs to be in place to ensure that potential exposures during the deployment have been recorded as well as a system for monitoring the long-term health of the deployed individuals. Following military campaigns in Iraq and Afghanistan the UK has developed research into this area by following specific cohorts of military personnel at periodic intervals.

> **Box 16.2.** Impact of the evolving viral haemorrhagic fever outbreak in the scenario
>
> The evolving VHF outbreak in neighbouring countries is particularly important as this could affect not just the health of the deployed force but also have a logistical impact. If flight restrictions and country entry screening programmes are implemented, then deployment and recovery of personnel and equipment may be hampered. Additionally, medical evacuation of ill personnel may become difficult.
>
> This outbreak will also mean further consideration of each of the sections already discussed. Some of the key points to consider are as follows:
>
> - Liaison with the World Health Organization and other bodies that will have detailed information on the VHF outbreak and its likelihood of spreading to the country that the task force has been asked to support.
> - The requirement for additional training of personnel in raising awareness of VHF and the specific preventive measures required to protect the task force against contracting the disease.
> - Provision of additional PPE. Service personnel historically were well trained in using PPE because of the threat of working in CBRN (chemical, biological, radiation and nuclear) environments. If VHF were to spread to the area where the deployed force is working there would be a potential risk if individuals came in to contact with infected patients or dead bodies.
> - The outbreak may lead to additional displacement of people from affected countries, which in turn may make the security situation worse.
> - Finally, if the VHF outbreak were to spread, then the force might see sights that had not been anticipated, which may cause or exacerbate any acute stress reaction that they may develop.

16.5 Lessons for Civilian Organizations

Military resilience when reacting to an emergency is underpinned by good planning and preparation of the force before it deploys. Therefore, any civilian organization that is likely to respond to national or international emergencies should ensure that it has done its 'homework' first before putting individuals on the ground. In essence, this means:

- understanding the area that the organization is going to deploy to and the local threats;
- having individuals who are 'fit for task' from personal medical, training, equipment and preventive perspectives;
- having a medical support and evacuation plan for deployed individuals; and
- the ability to protect the deployed individuals against security threats.

In any situation the perceived risks may change, as indeed may the tasks. Therefore, there must be a mechanism to constantly re-evaluate threats. This will require the organization to be agile and have sufficient backup capability to respond to changing circumstances.

Possibly the greatest risk for civilians who deploy in such situations is their return to home and normality. Organizations should therefore ensure they have systems in place to support individuals who may experience ill health when they reflect on the work that they carried out and the scenes that they saw. This should be proactive rather than reactive. It is recommended that in the early stages after deployment regular contact with individuals should be made and an assessment made as to whether they have successfully reintegrated into society. In the longer term some form of health surveillance of team members should be routinely planned. Consideration could be given to a pooling of resources post deployment in order that smaller NGOs could access the same level of support that larger ones may have. Taking this further, the United Nations High Commissioner for Refugees could play a role in delivering a post-deployment health surveillance programme.

16.6 Summary

Applying the military model of force preparation, maintenance and recovery may be of value to civilian organizations that respond to emergency situations. The main tenant is to ensure that the individual responders are able to carry out their deployment unimpeded by health issues. This will require regular risk assessment, knowledge of the health threats and risks in the area that they are deploying to, and a system to monitor their health on their return from deployment. Thus the responder organization should take the view that all responders are on a continuum of care.

Key Answers

- Military planning processes consider the 'what if' element of situations and have contingency plans for reasonable events. Thorough research informs planning and response arrangements, and there is substantial and continued engagement with partners to ensure common understanding.
- The military prepares, maintains and recovers its forces in response to an emergency by preparing staff before a response through thorough research of the situation, supporting them during a response with appropriate equipment and infrastructure, and supporting them when a response is over. These are key aspects of military deployments that could be adopted by civilian organizations.
- The continuum of care encompasses the pre-deployment, maintenance and recovery phases of deployment, as well as the stages from civilian life through recruitment and training before deployment, and through discharge and post-discharge from the forces.

Further Reading

Bricknell, M.C.M. and Moore, G. (2014) Safety and security: Health risk management – a tool for planning force health protection. In: Ryan, J.M., Hopperus Buma, A.P.C.C., Beadling, C.W., Mozumder, A., Nott, D.M., Rich, N.M., *et al.* (eds) *Conflict and Catastrophe Medicine: A Practical Guide*, 3rd edn. Springer-Verlag, London, pp. 369–376.

Colthirst, P.M., Berg, R.G., Denicolo, P. and Simecek, J.W. (2013) Operational cost analysis of dental emergencies for deployed US Army personnel during operation Iraqi freedom. *Military Medicine* 178, 427–431.

Iversen, A.C. and Greenberg, N. (2009) Mental health of regular and reserve military veterans. *Advances in Psychiatric Treatment* 15, 100–106.

Ministry of Defence (2015) Chapter 4 – Medical Intelligence. In: *Allied Joint Publication AJP-4.10, Edition B, Version 1, Allied Joint Doctrine for Medical Support with UK National Elements.* Available at: www.gov.uk/government/uploads/system/uploads/attachment_data/file/457142/20150824-AJP_4_10_med_spt_uk.pdf (accessed 3 March 2016).

Ministry of Defence (2015) Chapter 5 –Force Health Protection. In: *Allied Joint Publication AJP-4.10, Edition B, Version 1, Allied Joint Doctrine for Medical Support with UK National Elements.* Available at: www.gov.uk/government/uploads/system/uploads/attachment_data/file/457142/20150824-AJP_4_10_med_spt_uk.pdf (accessed 3 March 2016).

Rose, S.C., Bisson, J., Churchill, R. and Wessely, S. (2002) Psychological debriefing for preventing post traumatic stress disorder (PTSD). *Cochrane Database of Systematic Reviews* 2, CD000560.

Ross, D.A. (2012) Preventive medicine in the 21st century – a population challenge. *Journal of the Royal Army Med Corps* 158, 77–78.

Williams, R., Kemp, V.J. and Alexander, D.A. (2014) The psychosocial and mental health of people who are affected by conflict, catastrophes, terrorism, conflict and displacement. In: Ryan, J.M., Hopperus Buma, A.P.C.C., Beadling, C.W., Mozumder, A., Nott, D.M., Rich, N.M., *et al.* (eds) *Conflict and Catastrophe Medicine: A Practical Guide*, 3rd edn. Springer-Verlag, London, pp. 805–849.

Winslow, D. (2002) Strange bedfellows: NGOs and the military in humanitarian crisis. *International Journal of Peace Studies* 7(2), 35–55.

17 From Pandemics to Earthquakes: Health and Emergencies in Canterbury, New Zealand

A. HUMPHREY

Canterbury Medical Officer of Health, Christchurch, New Zealand

Key Questions

- What is the role of public health agencies in risk awareness, readiness, response and recovery from disasters?
- How does health link with other government agencies and the private sector (including the media) with respect to emergency planning?
- What specific health and public health issues could be expected in any emergency, so that preparation for one emergency can prepare a community for another?
- After a natural disaster, how long will recovery take and what kind of problems will health agencies be faced with in the recovery phase?

17.1 Introduction

The Canterbury earthquakes began on 4 September 2010 at 04.35 hours, when the Greendale Fault ruptured for the first time in 16,000 years. There followed more than 4000 aftershocks greater than 3.0 on the Richter scale, the most serious of which was a geoseismic 'punch' from immediately below the city of Christchurch on 22 February 2011. This earthquake recorded only 6.3 on the Richter scale but the shallowness of the earthquake gave rise to a peak ground acceleration of $2.2g$ and a modified Mercalli score of 10 (maximum intensity) in parts of the city. Overall, there were 185 deaths, most of which occurred in a single building that had been weakened by earlier earthquakes. Despite this relatively low absolute number by international standards, New Zealand has a small population and the Canterbury earthquakes produced the third highest national mortality rate from disasters in 2011, after Japan and Namibia, and the highest cost of disasters that year, as a percentage of gross domestic product (GDP).

While trauma and hospital care was important in the immediate aftermath of the February earthquake, water and sanitary health issues persisted into the medium term, while dealing with mental health issues continued for years post disaster. Many years of emergency preparation, predominantly for pandemics, had stood Canterbury in good stead for the earthquakes. The determinants of health are an important consideration in the years following the disaster as the Canterbury community rebuilds and maintains its commitment to resilience.

17.2 Legal Aspects

It is important that emergency planning, response and recovery is built on a firm legislative foundation, and New Zealand revisited its emergency legislation with the Civil Defence and Emergency Management (CDEM) Act 2002. This Act was a milestone in emergency preparedness in New Zealand as it moved away from a 'response' paradigm, dominated by emergency services preparing to assist communities in distress, to a multi-agency risk-reduction approach. In particular, the Act states that it aims to:

> improve and promote the sustainable management of hazards ... in a way that contributes to the social, economic, cultural, and environmental well-being and safety of the public and also to the protection of property.

This complements the World Health Organization (WHO) definition of health as a state of 'complete physical, mental and social wellbeing, and not merely the absence of disease or injury'.

Under the oversight of Canterbury Group Civil Defence organization (the provincial section of the national agency), local government, police, fire, ambulance and public health agencies have worked together since 2002 to integrate their planning and develop multi-agency capability for an all-hazards response. Building community resilience was seen as a critical component of this cross-agency planning.

17.3 The Influence of Pandemic Influenza: Pandemic Preparedness

The looming spectre of pandemic influenza provided a focus for building community resilience, not least because a serious influenza pandemic could be expected to cripple many services that would normally provide assistance. In particular health services, which employ many women of childbearing age, could be expected to be disproportionally affected by a pandemic, as many staff would not only be exposed to infection at work, but would also be diverted to care for their families once an epidemic took hold. The possibility of an influenza pandemic was therefore not only an important issue for emergency services to prepare for, but also focused all agencies on building community resilience. The message for the Canterbury community was: 'In the event of a pandemic, do not expect assistance from government agencies – they are likely to be more adversely affected than the general population'. In other words: 'It will be you and your neighbours who will succeed or fail fighting a pandemic – do not be dependent on the authorities to save you!'

Another interesting aspect to a pandemic emergency was the potential for health to be the lead agency for the response, a situation rarely experienced by police, fire, the military or others who generally regard health agencies as a small part of an overall response that is typically dominated by 'men in boots'. The Canterbury Group Civil Defence recognized the importance of health being able to take a lead in a pandemic and tasked the various health sectors (primary, secondary and public health) with delivering a cohesive plan and exercising that plan as a lead agency. In Canterbury it was a primary care organization of family doctors (general practitioners) who took the lead role, supported by public health and secondary care. The primary care organization hosted monthly meetings of all agencies for more than three years before they were put to the test by the emergence of A/H1N1 (Chapter 14, this volume). In addition, several exercises were conducted involving government agencies, communities and the private sector.

The monthly meetings were led by a triumvirate of the Chief Executive and Chair of the Christchurch Independent Practitioner's Association (a family doctor), the Canterbury Medical Officer of Health (a public health physician with specific legislative responsibilities and powers) and the Chief Medical Officer of Canterbury District Health Board (a senior clinical specialist who provides a clinical lead to hospital doctors in the region). All health agencies attended the meetings, including health service management and communications, representatives of the ambulance services, community pharmacists, laboratory services, district nurses, private hospitals, midwives, and other primary care organizations. In addition, Canterbury Group Civil Defence attended the meetings, providing oversight. Key to the response was the engagement of local and national media who were tasked with assisting in the planning and response, rather than simply being given media releases. Their positive engagement was to prove essential. Finally, private industry was also involved where needed. Agencies included local hotels (for isolation and potentially quarantine), bus services (for transporting people) and funeral directors. In addition, information sessions were delivered to a number of different agencies including the local business round table, schools, various levels of local government, and a variety of community groups including indigenous groups at local *marae* (Māori meeting houses).

Building community resilience had been an important part of the emergency planning strategy in Canterbury for several years. This was no less important with respect to pandemic awareness and preparation. The Ministry of Civil Defence and Emergency Management developed an information programme called 'Get Ready Get Thru' which provided useful information about food, water and emergency kit storage for a range of disasters as well as 'what to do' (Fig. 17.1).

(a)

GET READY GET THRU — CIVIL DEFENCE

HOW TO GET READY...

1. Learn about the disasters that can affect you
2. Create and practice a household emergency plan
3. Assemble and maintain emergency survival items
4. Have a getaway kit in case you have to leave in a hurry

Fig. 17.1. (a) Front page of a 'Get Ready Get Thru' leaflet (2010).

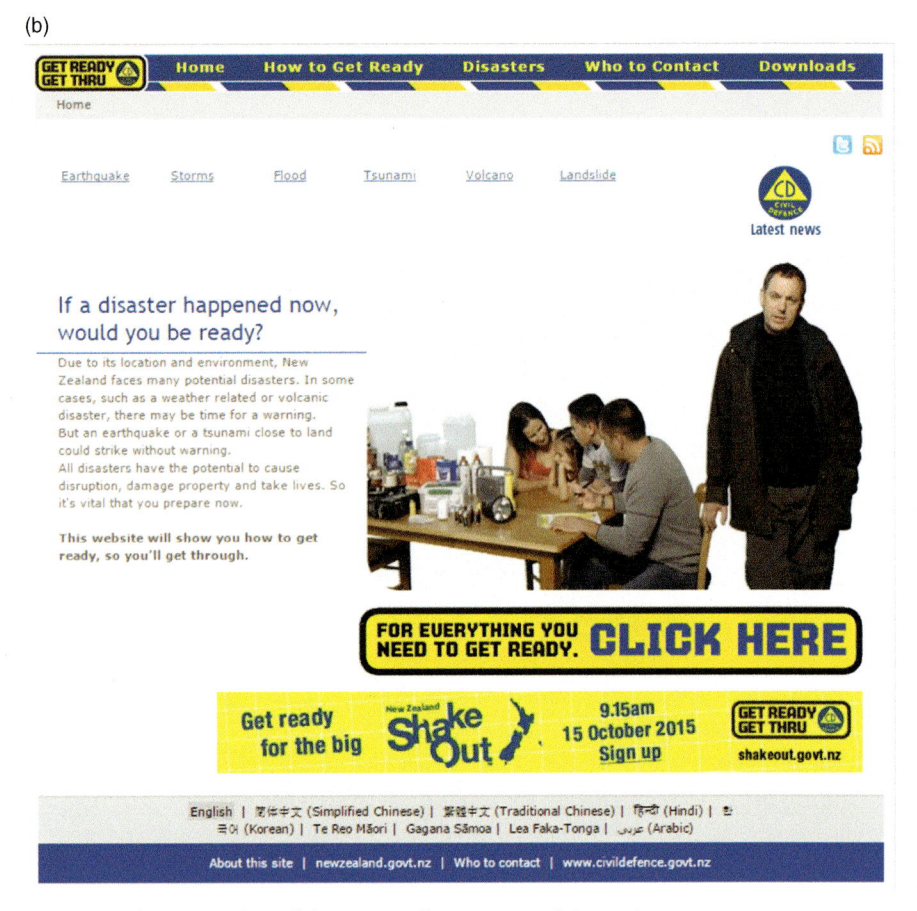

Fig. 17.1. (b) Screenshot of the current home page of the website (2015).

More specifically, the local science museum (Science Alive) worked with public health, virology and civil defence to develop a mobile exhibition called 'The Pandemic Survival Roadshow' which emphasized Cough etiquette, Hand hygiene, Isolation, Reducing germs and Preparation under the mnemonic 'CHIRP'. This roadshow was used to educate local government, schools, community groups and others over 2 years throughout Canterbury. The emphasis was on an isolated family surviving a week. Simultaneously, television and print media produced informative articles. Monitoring by the health communications team revealed that where journalists were engaged in the planning process, the information delivered by their stories was more informative and less sensationalized. In areas where journalists were not closely engaged, their stories about pandemics tended towards 'shroud waving' ('We are all going to die!') or conspiracy theory ('It's all a plot to sell more antivirals!'). Civil defence conducts national reviews on preparedness and in 2008 found that more than twice as many Cantabrians were aware of the pandemic risk, compared with the national average, and were also significantly more likely to be prepared for any form of emergency.

17.4 Pandemic Response

New Zealand was one of the first countries affected by A/H1N1 in 2009, when a party of high-school students returned from a school trip to Mexico via Los Angeles on flight NZ1 on 25 April 2009. An astute general practitioner in Auckland recognized the symptoms in a sick student and alerted the public health authorities to the possibility of 'Mexican swine flu'. Public health units across the country then embarked on a cluster control operation to identify, isolate, treat or prophylax (where necessary) more than 250 passengers from flight NZ1. There were 11 cases on the flight, two of whom contracted the illness during their journey.

New Zealand had a simple pandemic plan: 'Keep it out, Stamp it out, Manage it, Recover from it'. Although the use of quarantine for A/H1N1 was not advocated by the WHO at the time, New Zealand's relative isolation made quarantine a practical and likely effective option. There was a general expectation that 'Keep it out' (using quarantine) and 'Stamp it out' (cluster control) would be difficult to achieve and the country would steadily progress to 'Manage it'. However, the first cluster of cases was effectively isolated then treated, resulting in New Zealand possibly being the only country to completely 'stamp out' A/H1N1 in 2009 – at least for a while. Ultimately the disease returned, but stamping out the initial cluster bought 2 months of valuable time for health authorities and the community to get prepared.

17.4.1 Communications

During this period clear spokespeople were identified, such as the Medical Officer of Health, and the messages delivered by the pandemic roadshow over the last few years were credibly emphasized (Fig. 17.2). By the time case numbers of A/H1N1 started to climb again the community was ready to respond with self-isolation, appropriate social distancing and, above all, regular and thorough hand washing.

17.4.2 Dealing with tourists and travellers

In Canterbury, health authorities had worked closely with hotels for more than a year in preparation for a pandemic. Hotels were ready to cohort infected travellers, and others were ready to receive quarantined travellers. Health staff had provided infection control training to hotel housekeeping staff. Not only did this lift the confidence of employees (many of whom were new migrants) who had initially been anxious about turning up to work in a pandemic, but also provided valuable benefits after the pandemic when hotels and their staff became far more adept at dealing with other outbreaks. While local residents could be effectively isolated at home, travellers who had no local address were quarantined in a comfortable hotel, paid for by the

Fig. 17.2. 'The Pandemic Survival Roadshow' on public display (© A. Humphrey).

health authorities and visited at least daily by health staff. Most were happy with their (free) accommodation and all were compliant.

17.4.3 The main response

Cases could not be kept out indefinitely and once A/H1N1 became established in Canterbury fever clinics were set up in order to protect general practice and optimize infection control. A major benefit of the fever clinics (known as 'flu centres') was protection of the hospital system by effective triage from experienced general practitioners. At no stage did the hospital become overwhelmed with influenza, as it can do even during some 'normal' winter seasons. In addition, capacity to deliver routine care was maintained in the primary health care sector. In some areas of New Zealand, where fever clinics were not set up, primary care capacity became stretched, particularly when health care staff started to succumb to A/H1N1. Constructing, supplying and staffing the clinics were made possible through a clear budget assigned to the pandemic response by the Canterbury District Health Board. Not all areas of the country had such a clear mandate to act and costs were less clearly defined, with clinics and hospitals simply absorbing the workload.

Although A/H1N1 did not turn out to be as deadly as was predicted in April 2009, when the severity of notified Mexican cases artificially elevated the mortality rate, the overall response to the pandemic in New Zealand provided some useful lessons in emergency preparedness at both a provincial and a national level. Health and emergency planners in Canterbury were satisfied that their planning had contributed to flattening the epidemic curve and no service was overwhelmed at any time in 2009.

The successful aspects of Canterbury's response were threefold:

1. Clear information, delivered in an imaginative and compelling way, by recognizable and trusted spokespeople, with media agencies participating closely in the development and delivery of messages, rather than simply responding to the authorities' media releases.
2. A collaborative response, between different health providers, different agencies, with communities, the media and the private sector.
3. A clear budget for the response, clear delineation of responsibilities and the recognition of health agencies as emergency response leaders where appropriate.

These three broad areas were underpinned by detailed planning and trust. The conventional emergency responders trusted health to take the lead, all agencies trusted the media to act responsibly, and there was mutual trust between government agencies and the communities they served. The lessons learned from the A/H1N1 pandemic and the way in which Canterbury responded to this 'health' emergency stood the province in good stead when, a year later, a second emergency struck.

17.5 Transferring Learning from Pandemic to Earthquake Response

When the Greendale Fault ruptured at Darfield, west of Christchurch on 4 September 2010 at 04.35 hours, it produced a 7.3 magnitude earthquake in a region of New Zealand where an earthquake was not expected. Although there were some power outages and water supplies were cut briefly, structural damage was limited mainly to brick buildings and chimneys. Significantly, there were no deaths and it appeared as if New Zealand's rigorously applied building codes had done their job.

Many of the staff of the agencies involved in the response were the same as those who had responded to the pandemic a year earlier. They had each other's phone numbers on speed dial and were familiar with each other's way of working. This meant that the emergency response was both efficient and adaptable. Neither the local government emergency operations centre nor the public health unit's building could be used and alternative sites were identified. The identification of 'safe sites' from which to run an earthquake response was to prove extremely valuable in the future. A few buildings and many chimneys were demolished over the next few months – aftershocks were felt, some quite large, but few members of the community realized that the epicentres of the aftershocks were moving erratically but incrementally eastwards towards the city of Christchurch. People spoke a lot of their lucky escapes: the teenager who was thrown out the first floor of a building as it collapsed; the man who narrowly escaped being buried by his chimney bricks as he leapt up to save his collection of bottles; the couple who had got up early to catch a plane and, in the stillness and darkness following the earthquake, heard all the church bells ringing as the steeples swayed back and forth.

17.5.1 February 2011

Then, on Tuesday 22 February 2011 at 12.51 pm, while many people were meeting friends and colleagues for lunch, an aftershock struck that was to have a devastating effect on the city. There were no church bells ringing this time, as every church in the central city collapsed completely. Although only 6.3 in magnitude, the earthquake epicentre was close to the city and shallow (5 km), generating an unprecedented peak ground acceleration of $2.2g$ and the maximum modified Mercalli score of 10 in some parts of the city. Fortunately, the initial shake lasted only 12 seconds; otherwise many more buildings would have collapsed. The New Zealand building code requires buildings with an intended 50-year lifespan to be able to withstand a one in 500-year seismic event – the 22 February 2011 earthquake was considered a one in 2500-year event (Fig. 17.3).

There were 182 deaths in the first 24 hours following the earthquake and ensuing aftershocks, 115 in a single building that had been weakened by the earlier earthquakes. Ultimately 185 people were to die as a direct result of the Canterbury earthquakes – a tiny figure by international standards, but as a per capita mortality it placed New Zealand third in the world in 2011, after Japan and Namibia. Moreover, the cost of recovery has ballooned from an initially estimated NZ$12 billion to more than NZ$20 billion, easily the most expensive disaster, as a proportion of GDP, in any developed country, ever.

17.5.2 Impact on health care

The main hospital survived the earthquakes but suffered much infrastructure damage. Power across the city was lost immediately and remained out

Fig. 17.3. Liquefaction dust rising above the 22-storey buildings in the city of Christchurch immediately post-earthquake (© Gillian Needham).

for most of the central business district for several days. Although hospital backup generators kicked in within seconds, they had acquired sludge in the diesel tanks, which blocked the fuel pipes, causing further blackouts during the day. Other infrastructural difficulties included gases, fuel supply, power, boilers, sewerage, water, electricity, pipelines and air-conditioning among many other issues that required running repairs to keep the hospital going. The hospital's large in-house maintenance and engineering staff were critical to their success under extreme duress – their local knowledge and ability to work together could never have occurred if the hospital had been reliant on contractors for its maintenance.

Many patients presented to the emergency department via a variety of modes of transport; many refused to enter the hospital as they were worried about aftershocks, so a treatment area had to be set up outside the hospital. Registration and tracking of patients was difficult with computer systems down, and communication was also difficult. There were 6659 injuries and 142 hospital admissions in the five days following the earthquake of which just over half occurred on the first day. Although there were four amputees (three bilateral), 10 fasciotomies (this is a limb-saving surgical procedure where the fascia is cut to relieve tension or pressure to treat loss of circulation to an area of tissue or muscle) and 14 patients with crush injury syndrome, the vast majority of injuries were minor.

17.5.3 Coordinating the response

Outside the hospital, emergency services established the emergency operations centre in the City Art Gallery, as they had done 5 months before, as it was a new building known to be earthquake resilient. A national emergency was declared on 23 February 2011 (New Zealand's first national civil defence emergency) with the Chief Executive of the Ministry of Civil Defence as national controller. The relationship of health to emergency planning was not as strong at a national level as it was in Canterbury, but health representation was included in the executive. The inclusion of health at a high level, rather than subsuming it as part of welfare, was critical to a successful response. Trauma, search and rescue efforts are critical in the first few days (some would say hours) after the initial event. Water, sanitation and the consequent health risks are important considerations for weeks, sometimes months after nearly any type of disaster.

17.5.4 The wider response: water, sanitation and sewerage

With more than 80% of the central city's water and sewerage system severely damaged, there was a high risk of enteric illness. Power was out across large parts of the city for three days, so although a boil water notice was issued people had to use gas-powered barbecues to boil their water. One lesson from the earlier earthquake was to encourage people only to bring their water to the boil – not to 'boil it for three minutes' as had previously been advised. Protracted boiling has no health benefit and simply reduces valuable gas supplies. Once power was restored, boiling water became easier, but it would be many weeks before some houses had mains (reticulated) water restored.

The scale of the damage to the mains water system was such that the available number of registered water supply tankers was inadequate. Milk tankers were therefore requisitioned to deliver water to neighbourhoods which had no water. These became a useful focal point for providing messages to communities, including boil water notices.

Sanitation was also an immediate issue. Some areas, including the hospital, had a limited but fragile water supply, and a water conservation message of 'If it's brown flush it down – if it's yellow let it mellow' was used to minimize water wastage. Large parts of the central and eastern city had no water at all for many weeks, and a number of solutions to no toilets were promoted. Information about digging backyard latrines was provided and morale was boosted with 'dunny competitions' that were published in local newspapers. One innovative way of sewage disposal was to use plastic bags and dispose of them, once used, in waterproof garbage bins. The waste could then be removed by leak-proof garbage trucks that delivered to a landfill accredited to take liquid waste. Ultimately, thousands of portaloos were brought into the city and positioned on the sides of streets for people to use – not ideal in the middle of the night for the elderly or debilitated, but better than a hole in the garden. More than 10,000 chemical toilets were provided for the elderly to use in their homes.

Drinking water had been contaminated by broken sewers adjacent to damaged drinking water reticulation. Extra monitoring of the city water supply was therefore needed and assistance was provided to the public health staff by defence force environmental health officers. The extent of *Escherichia coli* transgressions clearly indicated that it would be necessary to chlorinate Christchurch's usually un-chlorinated, aquifer-fed water supply. It took many weeks for the chlorine to reach an effective level, as biofilm had accumulated over many decades in the pipes and rapidly soaked up any chlorine pumped into the system.

Until the water was effectively chlorinated, the boil water notice had to stay in place. It is well recognized that boil water notices are rarely rigorously obeyed after the first week, but a survey 6 weeks after the earthquake indicated that the boil water notice was still being observed by 88% of the population all the time. As with the pandemic, hand washing was encouraged and alcohol-based hand cleaning gel was distributed to those areas of the city with no water.

The public health unit established a heightened gastroenteritis monitoring system using an augmented sentinel practice system normally used for detecting influenza during the winter. Despite many weeks without water or sanitation for large parts of the city, the surveillance did not detect a single gastroenteric outbreak in the months after the earthquake.

17.5.5 Communications

Communicating messages to the communities around the city was key to maintaining resilience and a variety of systems were used. As well as the usual media outlets, 'community briefings' were established in local parks, convened by the mayor with the participation of all the main services, including health. These meetings attracted many thousands of people, who were able to listen to and ask questions of leaders from the main emergency services, including public health. The meetings were also a useful vehicle for bringing together communities, who were able to share resources or identify those among them who needed special help.

17.5.6 Local citizen response

Using Facebook, a small group of local students recruited thousands of others to assist around the city doing odd jobs, carrying messages or shovelling liquefaction silt out of people's homes and driveways. This 'student army' had originally been discouraged by the professional emergency responders, but they persisted and were soon welcomed to the response. In addition, farmers from out of town, who were less severely affected in the February 2011 earthquake than they had been in the September 2010 quake, brought in heavy equipment. The 'farmy army' and the 'student army' were great examples of how home-grown, self-reliant sectors of the community can provide extremely valuable assistance to the 'experts'.

There were many other examples of communities helping themselves: tradesmen helped set up laundries in welfare centres; Māori from other *iwi* (tribes) came from around the country to assist local Māori providing a door-to-door service to the vulnerable around the city; sports clubs away from the affected areas provided free showers for people from areas with no water supply. Mutual support appeared to be the norm, and despite much being made in the media of isolated cases of looting, the police recorded a drop in reported burglaries of 38% following the February earthquake.

Notwithstanding what appeared to be relatively high levels of social capital in Christchurch, poverty and inequity were, inevitably, an issue in the aftermath of the earthquake. Wealthier communities appeared to be able to attract assistance better than poorer communities; one example was that more portaloos were delivered in middle-class areas compared with poorer areas. The poorer neighbourhoods of the city, as with many cities, were built on poorer-quality, low-lying land. As a consequence, these neighbourhoods suffered from devastating liquefaction and more sewage spills – ultimately leading to colder, damper and more crowded homes in the immediate and medium-term aftermath of the quakes. The Australian Medical Assistance Team (Ausmat) set up a field hospital in the poor, eastern area of the city and, in the absence of much trauma, spent a lot of time listening to the stories and complaints from local residents. Many Ausmat staff were concerned that social unrest would break out. Fortunately this did not eventuate; but population loss has been greatest from the poorest areas following the earthquakes – a sign not only of unrepaired damage in this area, but also of disillusionment and despair.

17.5.7 The medium- to longer-term response

As the months have turned to years following the earthquakes, the aftershocks have gradually died away and people are now far more psychologically traumatized by struggles with insurance companies or repairers. Housing has become a critical issue for Christchurch, with affordable housing becoming extremely rare as workers have moved in to assist with the rebuild and homes have been demolished. People with mental health issues who would normally have been discharged home for community care are increasingly functionally homeless, and the mental health facilities are now constantly working at capacity. Health agencies have assisted with well-being projects (including mental well-being) including community gardens, community cycle projects, food forests, and 'gap fillers' (where interesting art, gardens or activities are placed on vacant lots where buildings have been demolished). In addition, a major health-led project aimed at encouraging people to celebrate small achievements and delights, to seek help and discuss their problems with each other – the 'All Right?' campaign – has been delivered by the Canterbury District Health Board (Fig. 17.4). It may have helped, but in the absence of affordable housing it is not enough.

Although the number of people living in Christchurch city has declined slightly since the earthquakes, the overall population of the region has increased. The rebuild appears to have stimulated growth with unemployment

Fig. 17.4. The 'All Right?' campaign (© Aaron Campbell Photography).

half the national average. Nearly 900 central business district properties and more than 5000 residential properties were demolished in the 4 years following the first earthquake and the city is now embarking on an ambitious rebuild programme.

The Canterbury Earthquake Recovery Authority (CERA) was established by an order of the Executive Council of the New Zealand Government on 28 March 2011 with its terms of reference broadly defined by the Canterbury Earthquake Recovery Act, assented 1 month later. The purpose of the Act (and CERA itself) was to provide appropriate measures to enable a focused, timely and expedited recovery from the earthquakes with the participation of the Canterbury community. CERA is a central government department with its own Minister working in collaboration with local agencies. A public consultation called 'Share an Idea' was promulgated by Christchurch City Council immediately following the earthquakes and this fed into CERA's Central City Development Unit's City Plan. More than 106,000 individual ideas from 'Share an Idea' were collated and categorized. Strong themes emerged around active transport, air and water quality, culture and vibrancy. It closely mirrored the environmental determinants of health model of Barton *et al.* (Fig. 17.5).

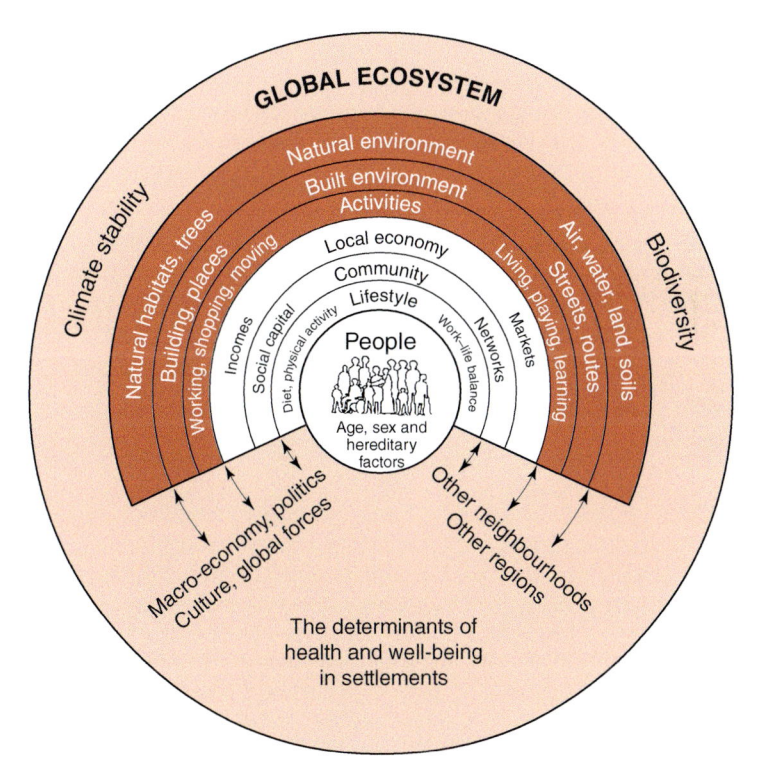

Fig. 17.5. Why public health is good at emergency preparedness: human ecology model of a settlement. (From Barton, H. and Grant, M. (2006) A health map for the local human habitat. *The Journal for the Royal Society for the Promotion of Health* 126 (6), 252–253. ISSN 1466-4240. http://dx.doi.org/10.1177/1466424006070466. Republished under the Creative Commons licence.)

As a consequence, the new city is planned to be low rise, with more green spaces, and will be centred on specific anchor projects such as a cultural precinct, a health precinct (including the hospital) and an innovation precinct for business development, among others. As a community Christchurch has centred its future firmly on well-being with wealth, growth, employment and other health determinants in a supportive role, rather than the *raison d'être* of the city. Clearly public health and public health agencies have an important supportive role in the city's aspirations. To that end, the New Zealand Public Health Unit produced an Integrated Recovery Guide based on a 'Health in all Policies' health determinants approach.

17.6 Summary

In conclusion, there were some valuable lessons learned during the influenza pandemic by all agencies about the important role of health and health agencies in all emergencies, not just so-called 'health emergencies', and in all aspects of emergency planning. Trauma is well recognized but once managed (and often this is quickly) water and sanitary health risks can take centre stage for many weeks or months. Mental health issues can last many years, and effective recovery from a disaster requires a community-based well-being centred approach – a key skill of public health practitioners. A review of the resilience literature revealed the five commonest themes associated with community resilience: (i) communication; (ii) learning; (iii) adaptability; (iv) risk awareness; and (v) social capital. Health agencies are well placed to develop these characteristics in the communities they serve, and in Canterbury, New Zealand health agencies played an integral role in developing these characteristics before and during the pandemic of 2009. The Canterbury earthquakes of 2010/11 demonstrated the importance of these characteristics in any disaster.

Ultimately there is no such thing as a 'health emergency'. All emergencies are health emergencies.

Key Answers

- Public health agencies have a key role in promoting risk awareness among the communities they serve, preparing those communities for emergencies by working with a range of government and non-government agencies, including the private sector. Public health is uniquely placed in the health sector to bring together diverse groups in a 'health in all policies' approach. Public health agencies also have an operational role in communicable disease management, border control, and water and sanitary health. In recovery, a 'determinants of health' approach is still important, with housing, employment and education issues impinging on health and mental health.
- Primary and secondary care's clinical focus does not lend itself well to interagency working. By contrast, public health agencies are used to working with local and central government, non-government organizations and the private sector, and therefore public health is the

Continued

> **Key Answers** Continued
>
> mortar that holds the bricks of emergency planning together. Risk communication is one area of public health expertise, but this is best delivered by working as a partner with local and national media.
> - A number of operational issues are common to any natural disaster, such as water and sanitary health, communicable disease control, surveillance and mental health issues. While risk awareness is important, an all-hazards approach rarely captures the imagination of the public in the way that a specifically publicized hazard does. However, it is important that the information and skills conveyed to a community can be transferable. These might include social connectedness, hand washing and infection control techniques.
> - Recovery from a major natural disaster can take many years, during which time the nature of the problems faced by a community will change. The incidence of mental health issues climbs during the recovery phase and is heavily determined by determinants such as housing, employment and education.

Further Reading

Ardagh, M.W., Richardson, S.K., Robinson, V., Than, M., Gee, P., Henderson, S., *et al.* (2012) The initial health-system response to the earthquake in Christchurch, New Zealand, in February, 2011. *Lancet* 379, 2109–2115.

Barton, H., Grant, M., Mitcham, C. and Tsouros, C. (2009) Healthy urban planning in European cities. *Health Promotion International* 24, i91–i99.

Castleden, M., McKee, M., Murray, V. and Leonardi, G. (2011) Resilience thinking in health protection. *Journal of Public Health* 33, 369–377.

New Zealand Public Health Clinical Network (2011) Core Public Health Functions for New Zealand. Available at: www.rph.org.nz/content/1a79cc74-2d88-45de-ba1b-31b301ade08f.cmr (accessed 8 October 2015).

Villamor, P., Barrell, D., Litchfield, N., Van Dissen, R., Hornblow, S. and Levick, S. (2011) Greendale Fault: investigation of surface rupture characteristics for fault avoidance zonation. GNS Science Consultancy Report 2011/121. Available at: http://ecan.govt.nz/publications/Reports/fault-final-report-greendale.pdf (accessed 8 October 2015).

Index

7/7 bombings (London, 2005) 41–42, 43, 46, 144

access to goods and equipment 46–47
access to mental health care/social care 94
accountability 127
Accountable Emergency Officers (AEOs) 14, 40
activation (triggers) of emergency plans 32, 76, 169–170, 179
After Action Review process 132
all-hazards approach 169
altruism 84
Ambulance Incident Commanders (AICs) 142
ambulance service 30–31
　in CBRN incidents 172
　in MCIs 141–142, 144, 147
antibiotics 162
anti-task behaviours 52, 53, 55
antiviral drugs 160
anxiety 52–53
audit of business continuity plans 100, 102, 106
avian influenza 156, 160

basic assumption behaviour 52
big bang incidents 4, 61
biological incidents 168, 170–171
　bioterrorism 163
　see also CBRN (chemical, biological, radiation, nuclear) incidents
bird flu 156, 160
bleach 176
Bloom's taxonomy 112, 113
bow tie model 22, 23
British Red Cross (BRC) 45–46, 47

'bronze' (operational) command 65–66, 69, 112
business continuity management 4, 99–108, 110
business impact assessment (BIA) 102–103

C3 (command, control and communication) 33, 60–71
　Christchurch earthquake 203
　command structures 64–69
　communication 61, 69–70, 116
　CONOPS documents 63–64
　control rooms 70, 122
　training and exercises 111–112, 116
Cabinet Office Briefing Room (COBR) 147
Cabinet Office National Risk Register 3
Canterbury (New Zealand)
　earthquakes (2010/11) 94, 194–195, 201–208
　pandemic preparedness 195–201
Casualty Bureaus 47
casualty clearing stations 142
CBRN (chemical, biological, radiation, nuclear) incidents 166–180
　planning 167–169, 171
　response 169–179
chemical incidents 167, 168, 169–170
Christchurch (New Zealand) earthquakes (2010/11) 94, 194–195, 201–208
Civil Contingencies Act (2004) (UK) 2, 38–39, 62, 79, 114, 128, 145–146
Civil Contingencies Secretariat (UK) 75, 77
Civil Defence and Emergency Management Act (2002) (New Zealand) 195
clothing removal 172–174
COBR (Cabinet Office Briefing Room) 147

211

code words used in exercises 123
command and control 33, 60–71
    Christchurch earthquake 203
    command structures 64–69
    communication 61, 69–70, 116
    CONOPS documents 63–64
    control rooms 70, 122
    training and exercises 111–112, 116
command post exercises 116
communication as part of the C3
    structure 61, 69–70
    exercises 116
communication with the public 33, 72–80, 201
    after an earthquake 205
    concerning infectious diseases 72–73,
        76, 162, 196–198, 199
community support/resilience 89, 90, 91, 94
    after the Christchurch
        earthquake 205–208
    in pandemics 195–196
computer failures 105, 106, 189
concept of operations (CONOPS)
    documents 63–64
conflicts of interest 53–55
consistency of information 76, 79
contact tracing 170, 199
continuity *see* business continuity
    management
continuum of care in the military 181–182
    before deployment 182–187
    deployment 187–190
    recovery/post-deployment 190
control 33, 61
    see also command and control
control of risk 26
control rooms 70
    EXCONs 119, 122
coordination 33, 66–69
    Christchurch earthquake 203
    in MCIs 145–146
corporate sector involvement 46–47, 196
costs
    of controlling risk 26
    of exercises 117
    in the recovery phase 149, 202
cybercrime 189

deaths
    in CBRN incidents 168, 177–178
    in the Christchurch earthquake 194, 202
    from influenza 104, 153, 155, 160
    in MCIs 139, 140–141, 145, 149
    treatment of the remains 104, 145, 177–178
debriefing 33, 123–124, 131–132
    see also lessons learned

decision making *see* command and control
decontamination 172, 174–177
    of the dead 177
defence in depth 157
definitions
    C3 61
    'emergency' 2
    incident types 4, 12, 61
    mass casualty incidents 139
    psychodynamic and systems theories 51
    psychosocial and mental health care
        83, 88
    in risk assessment 25
denial 52
dental disease 188
displaced persons 139, 145
disrobing 172–174
distress 86, 89
documentation of actions during an incident
    33, 68, 69–70, 149–150
drinking water 203–204
dry decontamination 176–177

earthquakes
    Haiti (2010) 140
    Nepal (2015) 140
    New Zealand (2010/11) 94, 194–195,
        201–208
Ebola outbreak 12, 45, 72–73, 163–164
emerging diseases 5–6, 12, 45, 72–73,
    163–164
ethics 162, 186
European Union (EU) 166–167
evacuation
    in CBRN incidents 171–172
    international 145, 189
evaluation of an exercise 121, 123–124
exercise control rooms (EXCON) 119, 122
exercises 15–16, 114–124
    business continuity 106
    for the London Olympics 63
experts
    in exercises 121
    in MCIs (STACs) 146
extreme weather events 14, 46, 139

facilitators (in exercises) 121
fatalities *see* deaths
fever clinics 200
field exercises 116, 117
field hospitals 144
finance *see* costs
fire and rescue service 144
flow charts in risk assessment 22

food contamination 163, 168
force protection (FP)/force health protection (FHP) 184
Foreign and Commonwealth Office (UK) 47
four-wheel drive cars 46

gap analysis 35
generic/core plans 32–33
geographical boundaries, working across 42–43
'Get Ready Get Thru' campaign 196–198
global planning 6, 7
global surveillance of influenza 157
'gold' (strategic) command 64–65, 69, 111
grounded theory 133

Haiti earthquake (2010) 140
hazard, definition 12
hazard and threat analysis 3
Hazardous Area Response Teams (HART) 144, 172
health registers 170
hidden agendas (anti-task behaviours) 52, 53, 55
Hillsborough disaster (1989) 149
horizon scanning 5, 68, 69
hospitals 4
  in the Christchurch earthquake 202–203
  in MCIs 142, 147–148
  in pandemics 158, 200
  as the scene of major incidents 4, 66
housing, after the Christchurch earthquake 206–207

immunization *see* vaccination
indexes, useful 32
industrial accidents 168
infection control 73, 158, 199
infectious diseases
  emerging diseases/Ebola 5–6, 12, 45, 72–73, 163–164
  VHF case study 181, 186, 191
  see also influenza pandemics
influenza pandemics 152–162, 164
  death rates 104, 153, 155, 160
  epidemiology 154–155, 161
  event triggers 76, 199
  preparedness and response (New Zealand) 195–201
  preparedness and response (UK) 13, 157–162
  surveillance 155–157

information management
  communication with the public 33, 72–80, 196–198, 201, 205
  as part of the C3 structure 61, 69–70
infrastructure damage in earthquakes 202–203, 203–204
interagency working *see* multi-agency working
international evacuation of patients 145, 189
international incidents 47, 76
International Red Cross/Red Crescent 44–46
  communications tools 74–75
isomorphic learning 110
IT attacks 189
IT failures 105, 106

jargon, avoiding 77
Joint Emergency Services Interoperability Programme (JESIP) 44, 65, 142

Kolb's cycle of learning 113

ladder pipe decontamination system 174
layered containment/mitigation 157
legal liability 16, 107, 149–150
legislation
  European cross-border health threats 166–167
  New Zealand 195
  UK (Civil Contingencies Act) 2, 38–39, 62, 79, 114, 128, 145–146
LESLP (London Emergency Services Liaison Panel) 44, 65
lessons learned 16, 33, 126–136
  business continuity plans 106–107
  isomorphic learning 110
  in MCIs 149, 150
  multi-agency team review 56
  post-exercise reports 123–124
live exercises 116, 117
Local Health Resilience Partnerships (LHRPs) 40–41, 112
Local Resilience Forums (LRFs) 39–40, 43–44, 112, 128
London 7/7 bombings (2005) 41–42, 43, 46, 144
London 2012 Olympic Games 60–61, 62, 63–64, 65, 66, 67, 69
London Emergency Services Liaison Panel (LESLP) 44, 65
London Resilience Team (LRT) 39
looking forward (horizon scanning) 5, 68, 69
LRFs (Local Resilience Forums) 39–40

Major Incident Medical Management and
    Support 143
mass casualty incidents (MCIs) 138–150
    Christchurch earthquake 202–203
    coordination 145–146
    London 7/7 bombings (2005) 41–42, 43,
        46, 144
    Mumbai (2008) 148
    recovery phase 148–150
    scope 138–141
    treatment of casualties 141–145,
        147–148, 203
    see also CBRN (chemical, biological,
        radiation, nuclear) incidents
master events list for exercises 119
media involvement 80, 145, 196, 198, 201
    see also social media
Medical Incident Advisors (MIAs) 141–142, 147
medical intelligence 182–184
medical risk assessment 184–185
meetings
    good practice 55–56
    negative behaviour in 51–52, 53
    of SCGs 67–68
    venues 55, 201
mental health care 83, 86, 89, 90, 93, 94–96
    see also psychosocial impact
MERIT 142, 144, 147–148
MERS (Middle Eastern respiratory
    syndrome) 163
military deployments 181–182, 192
    before deployment 182–187
    deployment 187–190
    lessons for civilian organizations
        191–192
    recovery/post-deployment 190
    and a VHF outbreak 191
missing persons 47
Mobile Emergency Response Immediate
    Teams (MERIT) 142, 144, 147–148
mortality rates see deaths
mortuaries 104, 145, 178
multi-agency working 14–15, 37–48
    Category 1/Category 2 responders
        38–39, 128, 146
    Civil Contingencies Act (UK) 38, 62,
        128, 145–146
    in control rooms 70
    corporate sector 46–47, 196
    group dynamics and teamwork 50–57
    intra-agency boundaries 42–43
    LHRPs 40–41
    LRFs 39–40, 43–44
    in MCIs 144–145, 145–146
    between military and other groups
        189–190

    in New Zealand 195
    SCGs 66–68, 146
    trust 37, 43, 201
    voluntary sector 43–47, 144–145,
        146, 205
    writing a plan 34–35
Mumbai terrorist attack (2008) 148

'Natech' accidents 169
National Health Service (NHS) 4, 14, 40–41,
    100, 112, 129, 139
    see also hospitals
National Occupational Standards 114
National Olympic Coordination Centre
    (NOCC) 62
natural disasters 139, 140, 169
    Christchurch earthquake 94, 194–195,
        201–208
Nepal earthquake (2015) 140
nerve agents 169, 170
neuraminidase inhibitors 160
New Zealand
    earthquakes (2010/11) 94, 194–195,
        201–208
    pandemic preparedness 195–201
NGOs (non-government
    organizations) 189–190
    see also voluntary sector
NHS (National Health Service) 4, 14, 40–41,
    100, 112, 129, 139
    see also hospitals
non-emergency health care facilities 41
    for military deployments 188
    self-reporting casualties 76, 179
nuclear incidents 149, 170–171
    see also CBRN (chemical, biological,
        radiation, nuclear) incidents

observers (of exercises) 121
Olympic Games (London 2012) 60–61, 62,
    63–64, 65, 66, 67, 69
Operation Amber 63
operational level command ('bronze') 65–66,
    69, 112
oseltamivir 160

pandemics 13
    influenza 76, 104, 152–162, 164, 195–201
    other diseases 163–164
    see also Ebola outbreak
panic 84
partnership engagement see multi-agency
    working

PFA (psychological first aid) 92, 94
planned events 34, 62, 138
    London 2012 Olympic Games 60–61, 62, 63–64, 65, 66, 67, 69
plans and planning 30–36
    benefits 1–2, 4–6
    for business continuity 4, 99–108
    for CBRN incidents 167–169, 171
    for communications disruption 79
    emergency planning cycle 10–17
    flexibility 5–6, 14, 31, 171
    gap analysis 35
    generic/core plans 32–33
    global 6, 7
    in the military 182–187
    multi-agency 33, 34–35
    for pandemics 13, 104, 157–164, 195–198
    for psychosocial care 92–96
    risk management 12, 18–28, 102
    structure of a plan 13–14, 31–32
    threat-specific 34
    validation 15–16, 26, 106
players (in exercises) 122
pneumonia 162
poisoning see CBRN (chemical, biological, radiation, nuclear) incidents
police service 40, 42, 67
post-incident communications 78, 205
post-incident follow-up 8, 33, 126–136
    business continuity plans 106–107
    Christchurch earthquake 206–208
    in MCIs 148–150
    psychosocial care 47, 83, 90–96, 206
post-incident psychosocial impact see psychosocial impact
post-traumatic stress disorder (PTSD) 85, 86, 93, 190
PPOSTT debriefing procedure 132
pre-hospital response to MCIs 141–145
primary task 51, 54–55
prioritization 13
protective equipment 171–172, 191
psychodynamic theory 51
psychological first aid (PFA) 92, 94
psychosocial impact 33, 82–97, 192
    of earthquakes 205–207
    of MCIs 149
    of military deployments 185–186, 188, 190
    psychosocial care 47, 83, 90–96, 206
    resilience 87–89, 94, 208
    on responders/staff 85, 92, 149
    stress/distress 84–87, 89–90
PTSD (post-traumatic stress disorder) 85, 86, 93, 190

public, communication with see communication with the public
public enquiries 16, 114, 127, 149–150
public health measures
    after the Christchurch earthquake 203–208
    pandemic influenza 157, 158–159, 196–200

quarantine 199–200

radiation incidents 149, 170–171
    see also CBRN (chemical, biological, radiation, nuclear) incidents
RASIO rate 188
record-keeping 33, 68, 69–70, 149–150
recovery phase 7, 33, 126–136
    Christchurch earthquake 206–208
    in MCIs 148–150
    in military deployments 190
    psychosocial care 47, 83, 90–96, 206
Red Cross/Red Crescent 44–46, 47
    communications tools 74–75
reflection/reflective practice 56
reputational risk/loss 5, 99, 117
resilience, psychosocial 87–89, 94, 208
responsibilities see roles and responsibilities
rising-tide events 4
    see also Ebola outbreak; influenza pandemics
risk appetite 25
risk communication 74, 79, 162
risk of conducting exercises 117
risk factors for psychosocial distress 89–90
risk management 12, 18–29, 68
    business continuity plans 102
    in the military 182–186, 190
risk matrices 24–25
risk registers 26
risk triggers 21, 22
roles and responsibilities
    of Category 1/Category 2 responders 39
    command structure 64–69
    in exercises 119–122
    senior management 14, 66–68, 114, 146
root cause analysis 107, 133
    bow tie model 22, 23
Royal College of Psychiatrists Occasional Paper 94 (2014) 83, 91–92, 96

sanitation (after an earthquake) 204
SARS (severe acute respiratory syndrome) 163

scenario assessment in the planning cycle 11–12
scenario development in exercises 118–119
scenario-based discussions 116
SCGs (Strategic Coordinating Groups) 66–68, 146
school closures 157, 158
Scientific and Technical Advisory Cell (STAC) 146
self-reporting casualties 76, 179, 203
Sendai Framework for Disaster Risk Reduction 6, 7
senior management 14, 114
    SCG 66–68, 146
sewage disposal (after an earthquake) 204
'silver' (tactical) command 65, 68–69, 111
situation reports (SITREPS) 69
smallpox 163
social media 76, 77, 162, 205
social support 89, 90, 91, 94
    after the Christchurch earthquake 205–208
    in pandemics 195–196
Spanish flu pandemic (1918/19) 154, 155
spontaneous incidents, definition 61
STAC (Scientific and Technical Advisory Cell) 146
staff
    business continuity plans for loss of staff 103
    control room 70
    in MCIs 144, 149
    in pandemics 195
    post-incident support 33, 85, 92, 149
    training *see* training
'STEPS 1-2-3 PLUS' procedure 179
stockpiling
    drugs 160, 162
    influenza vaccines 161–162
    smallpox vaccines 163
Strategic Coordinating Groups (SCGs) 66–68, 146
strategic level command ('gold') 64–65, 69, 111
stress 85–87, 89, 185
supermarkets 47
supply chain continuity 104
Support Lines 47
surveillance
    development of mental disorders 94
    influenza 155–157
    military personnel 188, 190
survivor reception centres 145
swine flu pandemic (2009/10) 76, 154, 155, 160, 161
    in New Zealand 199–201
systems theory 51

tabletop exercises 116
Tactical Coordination Groups (TCGs) 68–69, 146
tactical level command ('silver') 65, 68–69, 111
team leaders 53
    LRF/LHRP Chairs 40
    SCG Chairs 67
teamwork *see* multi-agency working
telephone helplines 47
terminology *see* definitions
terrorist incidents 141
    bioterrorism 163
    London 7/7 bombings (2005) 41–42, 43, 46, 144
    Mumbai (2008) 148
    see also CBRN (chemical, biological, radiation, nuclear) incidents
threat, definition 12
threat and hazard analysis (UK) 3
threat-specific plans 34
timing
    CBRN incidents 170–171, 172
    communications 77, 78
    exercises 117
    psychosocial impact 89, 96
tooth disease 188
training 15, 110–114, 124
    business continuity 105–106, 110
    control room staff 70
    exercises 15–16, 106, 114–124
    military personnel 186–187
training needs analysis 110–111
transport in emergencies 46, 189
transportation incidents 68
trauma risk management 190
triggers
    for activation of plans 32, 76, 169–170, 179
    in CBRN incidents 169–170, 179
    for de-escalation 33
    risk triggers 21–22
trust 37, 43, 201

umpires (in exercises) 121
uncertainty, in communication 74, 77
United Nations, Sendai Framework 6, 7

vaccination
    influenza 153, 160–162
    military personnel 185
validation of plans 15–16, 27, 106
viral haemorrhagic fever (VHF) case study 181, 186, 191

voluntary sector 43–47
    after the Christchurch earthquake 205
    in MCIs 144–145, 146
Voluntary Sector Civil Protection Forum (VSCPF) 44
vulnerable people, communication with 79

water supply 203–204
water-based decontamination 173, 175–176

weather events 14, 46, 139
workshop exercises 116
World Health Organization (WHO)
    on communication strategies 72, 74, 77
    influenza surveillance 157
writing a plan 13–14, 30–36, 101–104

zanamivir 160